The Greening

DATE DUE

MAR - 9 2000	
MAR 1 5 2002	
APR 1 3 2002	

BRODART   Cat. No. 23-221

The Greening of Ethics

Richard Sylvan
and
David Bennett

The White Horse Press
Cambridge, UK

and

The University of Arizona Press
Tucson, USA

Copyright © Richard Sylvan and David Bennett 1994

First published 1994 by The White Horse Press, 10 High Street, Knapwell, Cambridge CB3 8NR, UK

Co-published in the USA by The University of Arizona Press, 1230 N. Park Avenue, Suite 102, Tucson, AZ 85719, USA

Set in 11-point Adobe Garamond
Printed and bound in Great Britain by Redwood Books, Melksham.
⊛ This book is printed on acid-free, archival-quality paper.

Library of Congress Cataloging-in-Publication Data

Sylvan, Richard.
    The greening of ethics / Richard Sylvan and David Bennett.
    Includes bibliographical references and index.
    ——p.——cm.
    ISBN 0-8165-1504-2 (acid-free paper)
    ISBN 0-8165-1529-8 (pbk.: acid-free paper)
    1. Environmental ethics. 2. Environmental ethics—Australia.
I. Bennett, David. 1945— . II. Title.
GE42.S95   1994   94-18432
179'.1—dc20   CIP

A catalogue record for this book is available from the British Library.

ISBN *British edition*
    1-874267-03-4 (cloth); 1-874267-04-9 (paper)

About the Authors

Richard Sylvan is one of Australia's most creative philosophers and a central figure in the field of environmental ethics. Though widely cited, much of his work in this field has to date been available only as locally produced discussion papers. His research interests include object theory, environmental philosophy and political theory.

David Bennett is presently an environment officer with the Aboriginal and Torres Strait Islander Commission in Canberra. He has been a teacher of philosophy, the director of the Mawson Graduate Centre for Environmental Studies at the University of Adelaide, and a professional actor.

Contents

Preface

Among the purposes of this book are these: to set out what environmental ethics is and is not, to investigate the product and its Australian production and development; for instance, to track main movements in environmental ethics in Australia, to show not only in what directions environmental ethics is proceeding but how it may be developed, to provide suggestions on how environmental ethics can reach a wider audience, and to recommend methods and actions for inculcating and promoting environmental ethics.

That at least is how we interpreted our UNESCO project. The present work, which went under the working title of *Damn Greenies*, grew out of a UNESCO study on environmental ethics in Australia, which we undertook in 1989. In order to explain the unusual character of certain chapters of the present work, such as those on ways and means and on development, we need to set out the UNESCO task. In final format it was to include 'the following items', suitably sandwiched between an introduction and a bibliography and appendices:

ii)   What are environmental ethics?

iii)   How environmental ethics may be developed?

iv)   Ways and means for inculcating environmental ethics at individual, family, community, national and global levels.

v)   Recommendations for action with respect to (iii) and (iv) above.

The later chapters, in particular, have been organized by starting out from these headings. Only much later did we uncover a new unifying theme, which we have superimposed, namely a greening of ethics and applications of this greening.

Considered in its larger setting, the project UNESCO kindly offered us made an environmental ethics look remarkably like an economic public good, a product a privatizing entrepreneur could however fashion, develop, and market. We have taken this conception of

1

an environmental ethic as a generally good good, as a focus of an intellectual business venture, and run with it – in two senses: we have played along with the guiding image, but we have also played to advantage with it. It has served to structure our whole presentation, as will become apparent.

(An important aside: most of this work is, or ought to be, easy to follow. Where harder patches appear, as perhaps in some of the next paragraphs, we suggest that any discouraged reader skip over offending paragraphs to succeeding easier material.)

The conception, of ethics as such a good, to the extent that it stands, has an entertaining corollary: Theoretical ethics, the study of ethics, ethical systems and notions, both analytic and descriptive, becomes a branch of economics, a result no doubt gratifying for economic imperialism, and accordingly for economic rationalism which includes take-over and asset-stripping of other social sciences. (But the result reverses the view from the now finally disintegrating philosophic empire, which accurately sees economics within the 'moral sciences' as based on a de*moral*ized utilitarianism. The reverse view is now reinforced through the new directive of sustainable development: because sustainability constraints are ethical, economics remains ethically governed. So, in brief, there is no neat boundary between ethics and economics, and there is an important two way flow of intellectual commodities.) Unfortunately, like too many public goods in these latter days of enthusiastic deregulation, ethics and their imperatives obtain pretty shoddy treatment under mainstream economics.

Intellectual goods such as ethics and ideologies differ in significant respects, however, from more standard goods, even normal public goods. For example, with normal goods, conventional economics supposes that the goal is to deliver optimal quantities. With an ethics that does not make very good sense.[1] While an entrepreneur can promote parts of an ethics, a buyer can only purchase for serious adoption at most one ethics. With more than one ethics there is a conflict, except insofar as the ethics are mere exhibits, not something adopted and lived with. Such differences begin to bring out the significant *disanalogies* between an ethic and a product. Like most analogies, the goods analogy for ethics is only so good. Here and there we shall have to stretch or even break it.

2

Pursuit of such comparisons also begins to disclose complexity and ambiguity in the notion of an ethic or ethics. Goods such as ethics may be singular or plural: an ethics (or ethics but not a goods) is, ethics (not ethic) are... . There is a latent ambiguity, which we shall suppose is resolved as follows. While ethics are no doubt plural – that is, there are many ethics, some better, some worse – still the term *ethics* can also be used as a generic singular, to signify the kind (or a select one of the kind). Thus environmental ethics, in their varying forms, shallow and deep, utilitarian or deontological, are all ethics; each constitutes an ethic. Some of these are intended to supply both a theory of morality and an advocated morality, features of the practice itself, and thus not only link practice (what people live by) and theory, but also bridge another possible ambiguity. For the term *ethic(s)* also signifies both

- an ethical system, a morality or substitute therefore or scheme for generating such, covering the whole of life, or specific parts of and activities within life (a 'normative ethic' from an insider's or adherent's viewpoint), and

- the science or study of some or all such systems or features thereof, analytic or descriptive (a 'metaethic' or 'descriptive ethic', depending on the investigation, from an outsider's stand-point).

As we subsequently explain in more detail, we follow the familiar practice of using the term *ethic(s)* in all of these senses, without restrictions which would exclude ethical systems that violate Kantian principles of morality (as do egoism and economism for instance); occasional confusion of senses can easily be resolved, by allusion to more specific types of ethics. *Environmental* ethic(s) spans a similar range of senses; but environmental ethics are of course ethics (in any of the senses) which take specific, and perhaps special, account of the (natural) environment and environmental and ecological issues. Most of these points, like many others in this compressed preface, are elaborated in the text.

The later structure of the work, on environmental education and action, follows the outline of the project UNESCO gave us (indeed towards the end we began by simply converting that brief outline into chapter headings). But further we have oriented the study towards – what

UNESCO apparently intended, in its original proposals, but did not finally specify – the Australian scene. The Australian scene almost affords a microcosm of world environmental ethics; virtually all the main positions are represented, though not all the rich variations on these positions. It is a scene that has not been portrayed in any detail since 1980; it is a scene that is substantially ignored in Northern publications.[2] It is a scene worth discovering and sharing.

It is also a scene undergoing rapid change since late 1989, the cut-off point for the original version of the project. For a new wave of Australian books on environmental ethics, ecofeminism, and green (mostly very faint green) economics and political theory is beginning to roll in and break. We shall take selective account (primarily in terms of availability and interest to us) of new material insofar as it is of ethical relevance; however we do not pretend to, what would be but temporary and is typically illusory, some sort of completeness of coverage.[3]

One of the features that makes this work different lies in its regional emphasis and ecoregional commitments. While we remain committed to percepts of regional philosophy (quite at variance with dominant dull academic philosophy in Australia), nevertheless we suppose that such regional philosophy has wider interest, some even terrestrial relevance and reach, and that Australian environmental philosophy has a significant contribution to offer, globally. Indeed, perhaps more surprisingly, some of us consider that Australia – because of its still extensively intact natural heritage, its comparative wealth, and its environmentally motivated, and evident, alternative sub-cultures – is so placed as to play a special role in terrestrial responses to major environmental problems and to the escalating global predicament. There is a multi-faceted predicament to face, more, an environmental crisis, as we shall subsequently demonstrate. In particular, the biosphere, as a system capable of supporting versatile and diverse life forms satisfactorily, will not tolerate indefinitely present patterns of energy and resource use, waste production and life-support-systems degradation, by concentrated human communities. Conditions for satisfactory lives for many species, including humankind, will deteriorate further in the next century, perhaps, disastrously, unless some fundamental changes are made, and made soon, to these patterns. Ideas and motivation for such fundamental changes, for an environmental transvaluation of widespread basic values, are accordingly needed (as arguments from action theory confirm), desperately needed. These can come from an ethically founded environmental turn, such as is already being worked out and applied in a few parts of Australia. Indeed there is a case worth making that, in the circumstances faced, they can only derive from such philosophical and lifestyle sources. We shall develop that case in what follows.

We much appreciate the comments we received on an earlier version of this exercise in regional philosophy from Ian Hinckfuss, Ian McAuley and Joseph Wayne Smith and on a late version from William Grey and Fabian Sack. We should like to acknowledge the production assistance of Frances Redrup, Debbie Trew and Ria van de Zandt, and subsidization of their work and one author by the Australian National University. Finally, we owe a heavy debt of gratitude to the publisher, Andrew Johnson, not merely for his enthusiastic encouragement of our work, and his endless patience, but for his extensive editorial assistance. To him, indirectly, we owe the new unifying theme of the work, the greening of ethics.

1

Ethics and its Reluctant Greening, Set against Escalating Environmental Problems

INITIAL REASONS FOR A GREENING OF ETHICS

The environment is not a luxury. When political movements have faded, when economic systems have changed, when ideologies have been superseded and forgotten, the environment will still be important. Nor is it a free good. Human creatures, like others, depend on a satisfactory environment for their well-being and their very survival. But in their dealings with it, so-called developed societies have learned hubris, not wisdom. Instead of seeking to co-operate with and integrate with the environment, they have attempted to control it, to dominate it and to separate themselves from it. In the conventional wisdom and overweening pride of dominant Western traditions, many reasons have been provided for treating the environment as a passive, inanimate object which can be manipulated without repercussion. Yet despite this pride and these reasons, a hard lesson is being learned. Even if civilized humans do not care what happens to other species and ecological systems, there are repercussions for humans, as the environment is wrecked. However, if humans do learn to care about what happens to other species and ecosystems – that is, to treat nature as if it mattered – then the repercussions to humans will be lessened. Hopefully, at the same time, humans will discover that they are one species among many and that they are inseparable from their environment. But more important, humans may also learn the lesson that they can no longer treat other species and the environment as mere objects created for the use of humans. The well-being of humans as well as that of other species and the environment as a whole depends on many more humans learning this.

Like all other animals, humans depend on the environment for satisfactory survival. What that environment is like, however, may now depend as much on human largesse as on ecological principles, and at this time, biological evolution in many regions already depends on both. Ecological principles, such as carrying capacities, food webs, intraspecific and interspecific competition and co-operation, place limits on how many and what species there may be. Species do not live in isolation from the environment and other species. No species is above or beyond or outside these principles and the existence of each species is related to and depends upon other species. Despite their hubris, humans cannot escape the necessity for resources to maintain their existence. But what they can try to do, what they thought they could do, is to modify the planet to support human needs without regard to the diversity of other life forms and of ecosystems. It is rather as if these particular needs were all that mattered, encompassed all value.

Throughout most of the brief history of Western philosophy, certain humans have been the sole objects of positive moral concern. The inclusion of the non-human world in the form of landscapes, non-human animal species, rivers, forests, or other natural items entered, if at all, only as the property of certain humans or an organization of humans, such as the state, or because they were of interest to humans. In other words, the non-human world did not qualify in and of itself as an object of moral concern or even as the sort of thing that could be considered for inclusion. The reasons for this were and are many, but in the past century and more specifically in the past three decades these reasons have come under increasing challenge. One of the prime challenges has come as a response to the 'environmental crisis'. We in Western societies have become increasingly aware, stage by escalating stage, that something is wrong, drastically wrong, with the way that we have used and abused the environment. The central issues of the environmental crisis, pollution, human overpopulation, environmental degradation, and the endangering or extinction of species, have forced Western societies to re-evaluate, if ever too little and generally too late, their relationships to the environment and their responsibility to or for nature. Increasingly the impact of human activities on the biosphere's

7

capacity to support human life became a concern. Contamination of the environment is proceeding at a pace faster than natural processes can neutralize or disperse it. The possibility of human fertility outstripping the capacity of natural resources to sustain our species has become very real. The increasing rapidity with which other species and their habitats were disappearing passed from a lamentable loss to a serious concern about an impoverishment of other species for human enjoyment, study and use, and, more important, for themselves.

Out of these issues and concerns, not least concern for human life, and quality of life, but not only these concerns, arose environmental ethics. The idea that the non-human world *could* be part of human-centred ethics is not new. Although he rejected the idea, Aristotle contemplated it. In the Bible, when God set the rainbow in the sky, he made a covenant with humans *and animals* that he would not destroy the world again by flood, and God remains both the producer and the 'owner' of all creation with humanity only God's stewards. "When we believe that God made the world for us alone, it is a great mistake."[1] What is new, in the West, is the serious contemplation of including the non-human world under the aegis of moral concern, or even more startlingly *doing so*, extending adequate ethical treatment to parts of it. What is also new are many of the suggestions for how this might be done. But there is much disagreement about these suggestions. Some philosophers have gone so far as to "call for a new ethic for the environment". Others have merely suggested that the environment or parts of it be included within current moral considerations. And still others have desired to find some middle ground that does more than extend current moral considerations and the incumbent problems entailed by them, and yet not to break too much new ground so that humans no longer are considered the items of greatest moral consideration.

There is fairly general agreement among environmental philosophers (also called, at least when further conditions are satisfied, *ecophilosophers*) that the environment should be looked at from an ethical perspective. They agree that environmental matters are important, that environmental matters have not in the past obtained adequate attention, and typically that ethics should play a significant role, a much larger role.

But that is where limited agreement tends to end. For among these same philosophers there are disagreements about what constitutes an environmental ethic, about how to achieve such an ethic, to what degree such a thing is desirable, and the extent to which philosophical resources are available to work needed changes in Western traditions.

To work past these disagreements, we shall go back to neglected basics, to *ethics* and its *environment*. Without a characterization of ethics, a satisfactory account of environmental ethics and its roles is remote.

An ethic indicates, among other things, appropriate and inappropriate behaviour and treatment and to whom it is applicable. An environmental ethic involves a reconceptualization of appropriate and inappropriate behaviour towards the environment, by humans, and applications of ethics directly to the environment. The term 'ethics' derives from the Greek 'ethos' meaning custom, character, people or system. According to the Oxford English Dictionary, 'ethos' has come to mean in English the characteristic spirit of a community. It also means the moral principles by which a person is guided. If an ethic includes both a sense of community and guiding moral principles, then a proper environmental ethic implies not only a redefining of treatment of the environment, but also suggests an inclusion of the environment in the wider moral community. As that inclusion involves practical as well as theoretical changes in human treatment of the environment, the principles for doing this are both theoretical and practical.

In Western traditions the characteristic spirit of community has been limited to humans or even to certain subsets of humans. Over half a century ago Albert Schweitzer commented, "The great fault of all ethics hitherto has been that they believed themselves to have to deal only with the relationships of man to man."[2] It is important to realize that a proper environmental ethic is more than correcting this 'fault' in ethics, more than just an expansion. An environmental ethic, as well as a revamping of ethics, is also an operational ethic. It includes recommendations for practical environmental action as well as expressions of principle. An environmental ethic includes both a description of what should be the case and particular suggestions for action. Now to improve upon these evocative preliminaries: –

AS TO ETHICS AND ITS PROPER CHARACTERIZATION; 'ENVIRONMENTAL ETHICS' AND ENVIRONMENTAL ETHICS

Even textbooks and guidebooks on ethics often make no attempt at a definition, and often have little or nothing of due generality to offer on what ethics is about.[3] Unremarkably then, many of those presenting and teaching ethics, perhaps well enough, have no adequate idea of what ethics is about. This is revealed by the sorts of definitions they slop down and by what they say ethics is about (the criticism includes what is said at the end of the previous section). The slop presented indicates they did not even bother to consult their dictionaries with any care.

A few sample definitions recently encountered will confirm our claims. Utterly defective is the account of ethics as 'the relation between you and other people'.[4] Such connections as the spatial relations between people, for instance where they sit at a seminar, do not belong to ethics. Narrowing relations to those of treatment, as in 'how you treat other people' improves the account a bit, but not enough. Decent treatment of relevant kinds no doubt comprises part of ethics, but a small part. Nor is such treatment confined, especially as regards receivers, to people.

A more interesting account, presumably designed to encompass applied ethics, takes this remarkable form: "'ethics' [is] defined as *disciplined reflection by persons in all walks of life on moral ideas and ideals*".[5] As a definition, this too fails entirely, appearing circular, and converting ethics away from, what is essential, practice, into head stuff, into something like meditation, a kind of widely practised theorizing, though practised on certain parts of ethics! There are too some unfortunate side-effects of the insertion, aimed at making ethics applicable to all walks of (human) life: namely, as there are walks of life which have no room for disciplined reflection, so there is in fact no ethics. Back again to the drawing board.

It is not difficult to do better than ethics teachers and writers have, simply by consulting a better dictionary. But it is also not so difficult to re-organize and improve upon what dictionaries offer. Like the preceding definition, the dictionaries seem mired in a common circularity: ethics is defined in terms of morality (and its adjuncts, decent treatment, and similar); but then morality is itself defined through ethics. Fortunately the loop back to ethics can be readily broken.

A particularly satisfactory way to workable definitions of notions in the *ethics* orbit – ethic, ethical, ethics, to take direct cases – is through those in the presupposed *moral* orbit – directly, moral, morality, morals. For all these ethical notions are regularly defined, in better dictionaries, through moral notions. This approach reverses the difficult and so far unsuccessful strategy, sometimes tried in ethical theory, of attempting to define morality (e.g. through universalizability conditions) inside a somehow independently characterized ethics.

The bases, upon which moral notions are carried, lie in turn in the *action* orbit, in features of actions and their agents.[6] As agents are entailed by actions – because actions just are agent-ascribed processes – actions, more generally processes, form bedrock here. Relevant elements of the action nexus, upon which moral actions are carried, are now indicated:

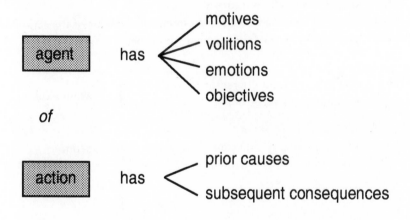

FIGURE 1.1: *Single action* snapshot, capturing
agent-acts-in-environment nexus

The breakdown of actions, in types , will straightaway yield much that has long been taken as central in moral theory, such as agonizing (as Socrates did) over choices. Choice-making is a morally critical type of action. *Plural actions* expose further relevant features, including conduct, dispositions and character of agents, and suggest typologies (of moral relevance) of actions and of impacts. Virtue ethics, but part of moral theory, fit here.

The sketchy action preliminaries admit of much elaboration and variation, both expansion and contraction, most of which however we can pick up as we proceed. For example, an illuminating expansion would expose the setting, of constraints and incentives, within which an agent acts; no doubt some of these features, incentives for instance, can be absorbed into what is already depicted, such as motives, and the remainder could be included under a prior-influences category. Striking contractions are those afforded by prevailing ethical theories, consequentialism and Kantianism, which would erase, respectively, all but consequences or all but motives.

The term *moral* covers the assessment of elements of the action nexus, as depicted, in the following sorts of respects (as applicable):

excellent, good, indifferent, bad, evil
valuable, of value, or not   } axiological

right or wrong (or alright), fair or unfair, proper or improper, obliga-tory, forbidden, permissible   } deontic

responsible, irresponsible   }

It also covers principles concerning such assessments.

Plainly some of the features are not significantly applicable to all elements; for instance, impacts are not significantly responsible. Respon-sibility primarily applies to agents, their conduct and character. The *sorts of respects* cited are more general forms. Many more specific evaluations are thereby encompassed; for instance, honourable, dishonourable; decent, indecent.

In terms of the sorts of respects, certain important subtypes can be distinguished. The first group (given by the first two lines) covers *axiological.* The second group (given by the next two lines) covers *deontic.* The third group (not usually distinguished in 'the logic of ethics') comprises *accountability* notions.

Such formerly important terms as *virtuous* and *vicious* – regularly, but wrongly included in dictionary listings of respects defining *moral* (e.g., "of or pertaining to character or disposition considered as ... virtuous or vicious") – are axiological. They can be characterized in terms of what is already available, thereby avoiding another circularity vexing dictionary definitions (which define 'moral' using 'virtuous', and 'virtuous' deploying 'moral'). For example, to adjust the dictionaries, virtuous [action] is [that] exhibiting moral excellence or quality, vicious that distinguished as morally evil. Much the same supervenient story gets told for *duty* in the deontic class. For while a duty just amounts to a certain sort, a rigorous sort, of moral obligation, duty is not among the respects listed in defining the term moral.

There are several, loosely consequential, attachments often made to what counts as moral(s), notably: issues of punishments and rewards, with a view to altering proper conduct; of freedom and determinism, as the latter may appear to place moral enterprise in jeopardy; of justification of moral principles, which do not stand on their own without moral justification; and so on, for other theoretical topics that consequently arise. While these further issues do not form part of dictionary definitions of *moral,* they are stock textbook fare in moral philosophy. But that, moral philosophy and moral theory, is where they can comfortably stay. While the issues that result could be called *moral,* or distinguished as textbook moral, they might as well simply be accommodated more accurately, within philosophy of morality (a part of meta-morals).

Morality is the attribute of being moral.[7] Thus an item exhibits morality if and only if that item is moral. In wider compass, morality comprehends a larger package concerning what is moral; it consists in the theory, practice and principles of what is moral. It follows, from the account of *moral* given, that morality is all about certain features of agency, features of theory and practice of agency, namely evaluative and assessment features and principles concerning these.

13

Several other significant corollaries also emerge. Firstly morality has nothing essentially to do with humans. A similar result will follow for ethics, which is characterized, also in a human independent way, by way of morality. Ethics too has nothing essentially to do with humans (or even with humanoids). Morality turns on features of agency. Humans simply happen to supply, presently, prime terrestrial examples of full moral agents, those actors who not merely conduct themselves well or badly (a kangaroo can be a good or poor mother) but who make promises, recognize obligations, accept responsibilities, and so forth. Humans were almost certainly not the only full terrestrial agents; other humanoid species that have become extinct (perhaps with human assistance) very likely were also. Competent humans are probably not the only full moral agents even now in this universe. It is unlikely that they will be the only moral agents, perhaps even terrestrially. The prominent role of competent humans in morality, and so as regards ethics, is utterly contingent. It follows that those familiar definitions of morality and ethics in terms of humans and their features are one and all defective. For all the green ethical theory to be developed, humans can drop out altogether. But competent humans are useful for illustrative purposes; it helps to have some demonstrably actual examples. And in practical terms of terrestrial ethical impact, they are now almost unavoidable.

Focusing upon morality *would* be unduly limiting from the perspective of environmental ethics. For although anthropocentricism has been avoided, easily circumvented through agency, though humans play no essential role, still morality is undoubtedly agent-centred. Morality foregrounds agents and assessments of them and their actions, and backgrounds environments of agents, and environments more generally (those without agents in them do not even enter into consideration). The environment remains, so to say, mere backdrop to actions of agents.

Before explaining how to escape actocentrism (agent-centredness) in ethics, it is worth recording that there are advantages in having actocentrism as a staging post. One advantage is that unwanted senses of evaluative notions are peeled off automatically, because they do not apply regards agency. For example, so removed are all those quasi-descriptive uses of commendatory terms such as *good* (e.g. 'good car', 'good sort', 'good sox', 'good ox').

Escape is through an intermediary, which is substituted in place of morality in the account proposed of ethics. That intermediary is an *expanded* morality, in convenient shorthand *ectomorality*.[8] ('Ectomorality' is close enough for the present to 'ecomorality', which is of course part of the main focus.) Expanded morality takes into account also *evaluative settings*, perhaps devoid of agents; whatever of value agents, or agent-like subjects, agentoids, might in principle impact upon, interfere with, disturb, affect, change, dislodge, or similar. That is, to say, it encompasses all items of relevant value, ethically *significant* items, including many wholes, such as habitats, ecosystems, the Earth itself; and so also whatever applies morally with respect to these, for instance permission to disturb or to improve, principles of letting be, and so on. Such ecosignificant items are also called, in the literature, morally *considerable* items (a nicely ambiguous term, both sides of which have unfortunate features, one because a considerable item may not have large significance, the other because what it is supposed to respond to is some restricting class of considerers). In yet other and more satisfactory terms, they have ethical *standing*. Naturally items of significance other than agentoids may have other ethically relevant features (which they perhaps accrue through their significance). For instance they may hold entitlements of various sorts, such as to be left alone, to be represented, and so on.

In order to characterize core notions in the ethics orbit, the basic strategy consists in plugging in 'expanded morality' where the dictionaries give 'morality'; substituting similarly for 'moral'. For the adjectival forms 'ethical' and (equivalently) 'ethic', that is all that is required. The form concerned means: pertaining to ectomorality, or (equivalently) treating of or relating to ectomorals. For the noun forms it is not so simple, because there are problems with the dictionary accounts (for instance in 'the science of morality' owing to misplaced definiteness, and dated uses of the term 'science'). Always *ethic(s)* is an abstract, signifying features of being ethical; more exactly, always concerning ectomoral theories or systems. More specifically, *ethic(s)* signifies

firstly, in both singular or plural, though usually plural, the whole field of ectomoral theory (the 'science' of ectomorals), or derivatively, the department of (philosophical) study concerned therewith;[9]

15

secondly, in the singular but pluralizing, a system or schematization of
ectomorality; that is, at least rudiments of a system, or of a
theory, comprising values, ectomoral principles, and rules of
proper conduct.

Like stock textbook ethics, most environmental ethics so far are in fact
truncated; they are very rudimentary schemes, leaving out much that
would be expected in a fatter systematization. Hence in part the
vagueness of characterization (which does not however exceed that of
dictionaries or of ordinary usage). So far the vagueness is immaterial as
the characterization is workable; but exactness can be obtained through
a technical upgrade of *system* and a more detailed specification of what
such a proposition system contains.[10]

What are sometimes cited as further restricted senses of 'ethic' are
not further senses. Rather they are restrictions, or designations, of what
is already presented. Thus for instance, the 'moral principles of a
particular leader or school of thought' as in Buddhist ethics or Kantian
ethics. Similarly for field indications or application restrictions, as in
medical ethics.

Something like the division between the first and second specifics
above has been quasi-technicalized within ethics, and distorted through
merger with a theory-metatheory distinction borrowed from outside
ethics. Engagement with the second (as in systematization, constructing
and working with systems, etc.) has been dubbed *normative* ethics, or
sometimes to increase the confusion just ethics, or condescendingly,
casuistry. *Metaethics* has been taken however to be not what the logical
analogy would suggest, meta-normative ethics, the (positivistically correct) study of such systems, but instead a brand of analytic philosophy,
analysing the meanings and uses of key ethical terms. Taken as exhaustive, there is much of importance, that the normative-metaethical
distinction leaves out. Nor is the distinction exhaustive. As well, the
labels are misleading. In short, the distinction is ill made out and
unsatisfactory; it can be abandoned without loss and to advantage. Let
us do so (so forget the last paragraph, if you prefer).

In order to reach an *adequate* definition of ethics, not merely for
intended environmental purposes, but for historical purposes as well (so
as to include Stoic ethics, Taoist ethics, romantic nature ethics and the

like as ethics) we have been obliged to *adjust* (if in an inconspicuous way) dictionary definitions of ethics. That adjustment was made through 'expanded morality', but could have been accomplished in other equivalent ways (Leopold's expansion of 'community', which we consider soon is another, less satisfactory, way of achieving the same sort of thing). Granted that we have adjusted dictionary definitions, have we changed the very notion of ethics, or did the dictionaries, not for the first item, get it wrong, getting caught with chauvinistic definitions? We claim that we have not changed the notion. Rather we captured what some meant, more or less, by ethics all along – Stoics, Epicureans and others. It has never been a very precise notion. In this case (by contrast with terms like *right*) not too much hangs on the term, and we could simply divide ethics up, for instance, into wide and narrow. However, we intend to keep the nice term ethics but to give the dictionaries and modern opposition the expressions, standard ethics, or conventional ethics.

Our inconspicuous but all important adjustment is both good and bad news. It is bad news, insofar as it appears to assist the charge that [deeper] environmental philosophy is doing something different from ethics, that it is guilty of changing the subject (i.e., effectively committing *ignoratio elenchi*). However we can freely grant, without incurring damage, that standard ethics was, as a matter of definition, at least actocentric, and that, insofar as agents were identified with humans, it was, therefore, anthropocentric. Even so ethics, as redefined, *includes* standard ethics, so it is more satisfactorily seen not as a different but a larger subject, expanded through an expanded morality. It is good news, not merely because it facilitates environmental ethics and accords with more satisfactory (greener) accounts of the subject, but because it helps in getting around a snarl of objections to the effect that ethics has to be homocentric, or at least actocentric. Removing these objections will recur throughout what follows.[11]

Issues in the characterization of ethics, especially removal of actocentrism and homocentrism, naturally get transmitted to critical types, such as environmental ethics. But it might be imagined, rashly, that disagreement ends there. After all, all environmental ethics make the non-human world a proper object of concern, either directly or indirectly. Differences between explications offered can be ascribed to ordinary philosophical incompleteness. "An environmental ethic, broadly, is a set of values to live by which takes due account of the value of the non-human world."[12] Stated differently, the aim of environmental ethics "is to understand and act appropriately in relation to our ecological – in the broad sense – circumstances".[13] Differently again, "...sets of principles, which would guide our treatment of wild nature, constitute an environmental ethic in the most general sense".[14] The first of these slack explanations leaves out what is essential for an ethics, principles; the second is unduly agent centred, leaving out what is not (such as agent independent values); the third not only neglects values, but excessively narrows scope so that such an ethics is inapplicable in places devoid of wild nature or to other issues. A better explication, which avoids these obvious pitfalls runs: An environmental ethics is an ethics, as defined, which among other concerns, takes specific account, through its values and principles, of (significant) environmental items, of parts of nonhuman nature. But it too leaves much to be explained. How the non-human world is made the object of concern, how it is moral concern, to what degree it is made the object of moral concern, what forms of moral concern are applicable to what elements of the environment, and how to achieve the aim of acting appropriately – all are matters that make environmental ethics one of the most controversial areas of philosophy and also one of the most exciting, although many professional philosophers still regard it as a peripheral area. Yet there now is broad agreement among most of those active in the area that some change is needed and that environmental matters need much more attention and action.

An environmental ethics is an ethics concerned, usually among other ethical things, with parts of the environment. Coal-face or factory-floor environmental ethics bears on environmental issues, problems and courses, and can be expected to say something about environments or some of their nonhuman components. Academic or board-room environmental ethics is frequently, however, a remove from direct ethical work, comprising discussion of environmental ethics, its features, its methods, its variability, its presumed merely derived and applied character, and so on. Much of this academic environmental ethics is directed *against* coal-face environmental ethics, especially that of a genuinely committed sort aimed at advancing environmental causes.

It is almost immediate that there is an important elasticity – ambiguity, in a slack sense – in the notion of *environmental ethics*, between an ethics which considers environmental matters, whether positively or negatively, and an ethics which is positive about some environmental elements, which evinces concern about environmental items (as 18th Century sympathy ethics did about other humans). A *superficial* environmental ethics need not say anything particularly positive or sympathetic about environments; indeed productions on the topic may try to dismiss or defeat environmental causes and to defuse or dissolve environmental problems, for instance as not worth bothering about.[15] It may be argued that natural environments should serve highest economic causes, which means in practice becoming reserves for city folks, city corporations and consumers. Under such environmental ethics ('environmental ethics' it is tempting to say, contravening usage) many an environment should be razed, levelled, and paved over, to become a grand parking lot, shopping mall or urban escarpment of a concrete jungle. What rural environment remains will, also serving those 'high' purposes, be tamed and managed, like much of Western Europe. Witness the enthusiasm in productions on environmental ethics for landscapes like that of Tuscany, where virtually everything is under tight human control and no bird moves (else it is shot).

Environmental ethics and environmental philosophy may well prove to be disappointments for enthusiastic environmentalists. For there may be little that is materially green about them, no commitment to changes in old attitudes and practices, no offers of improved standing and ethical treatment for environmental items. As an environmental economics may be no green economics, but (very likely) some part or application of mainstream economics, some sweep of resource or land economics, so an environmental ethics or environmental ethic may not be particularly or at all green, but, for instance, simply part of establishment ethics or an 'application' of it to environmental items. Fortunately not all environmental ethics is like this; beyond environmental is genuinely green ethics.

The elasticity in environmental ethics between superficial and material forms, parallels that of the term *concern* (and rather similarly of *matter*), and can be substantially constricted to that: an environmental ethics is one concerned with environmental items or matter. Thus, on the one construal, such an ethics may be simply about such topics, it may be a standard ethics applied to such topics, in which case environmental ethics does become just so-called applied ethics. Alternatively, however, such an ethic may be one evincing concern, worry, about the treatment of some environmental items or other, and directed at obtaining improved treatment. Such an ethic is not negative or indifferent towards environmental matters, but is advancing environmental causes.[16] As these matter with such ethics, environmental ethics cannot be an applied ethics; for the framework of standard ethics and their applications, is exceeded. (At best such ethics can be approximated by standard ethics by assuming enlightened agents holding appropriate sorts of values.[17])

An obvious method for surmounting this elasticity problem proceeds through distinctions. Let us distinguish *eco-ethics* (which will prove tantamount to *green* ethics) as those ethics positively concerned with, and not merely about, environmental topics. Then there is but an overlap between eco-ethics and environmentally applied standard ethics, *enviro-ethics* for short, since it is always standard ethics that obtain these applications. Such environmental ethics, of applied type, typically differ from standard ethics only in their explicit inclusion of or allusion to environmental elements. As regards their implications for the environment however, they may differ not at all; they may incorporate precisely the same theory and derived evaluations. Older standard ethics failed then to count as environmental only because they were silent on environmental matters, not because they lacked environmental impacts; they certainly did yield environmental results and evaluations.

If such enviro-ethics, standard applied ethics, were all there were to environmental ethics, then it would be a pretty sad affair. For all such ethics give environments much poorer treatment than they should be receiving, all unfairly twist evaluations in favour of agents, typically overrated humans. Indeed all such ethics are, for reasons building on these, defective.

Fortunately then, there are, as indicated, green ethics which are not such enviro-ethics, some of them rich and deep ethics. Virtually any of the deeper green ethics to be encountered later will illustrate the difference from merely applied ethics. Conversely there are, all too obviously, applied ethics which are not green ethics. Most ethics generated from the academies will obligingly exemplify. Green ethics, ethics that is that are green (in the familiar sense to be supplied), diverge from what applied ethics supplies, with the divergence increasing steeply with depth of greenness.

EXPLAINING GREEN AND THE REAL GREENING OF ETHICS

In characterizing *green* we start again with agency. Such relevant features about environmentalists, of greens, as being positively concerned about nature or natural things, supporting environmental causes and similar, are closely tied to practice and action; they are features of agents. In an original central sense of 'green' so redeployed, an agent was *green* if that agent firstly believed that some parts of the natural environment should be protected, and secondly took some relevant action.[18] In short, a belief-action analysis emerges, which can be recast as a standard belief-desire-resultant action account. For instance, the agent believes that some natural environment is worth protection, desires that it should be, and acts (in some way, perhaps feebly) accordingly. All the parts of this sort of account call, however, for some elaboration. To remove a certain circularity, resulting from characterizing 'green' through terms like 'environment', 'natural environment' can be replaced by 'parts (surface configurations) of the Earth substantially uninterfered with by humans'. To specify protection satisfactorily, protection *from* what should be indicated. And so on. But there is little point in elaborating this already weak characterization, which can be satisfied by an agent interested in saving a few square metres of the Earth's surface from exploitation. For the notion has been further weakened, almost from its inception.

To characterize this laxer, but ubiquitous notion, let us insert the intermediary notion of an *environmental cause*. Such causes include not merely *protecting* natural environments, but *limiting impacts* on environments, which may not be pristine, for instance by pouring less sewage or detergents, CFCs or Greenhouse gases, into them, or *conserving* items, which could be exploited or consumed now for future use or for future human generations (to use or conserve, etc.). The first of these further causes beyond protection, original green, is a largely consumerist notion of green, the second, an economistic notion (it takes in such beloved constraints as intergenerational equity). There are other related causes of importance also, such as *restoration* of something that has been over-exploited or damaged, such as agricultural lands, waterways, urban air, soils (soil conservation), or even human artefacts such as monuments or buildings. The shallow green cause of fighting pollution, for human health reasons, can fit in here. Needless to add the restoration may not be to pristine conditions, but to quite low standards. In sum, causes include as well as protection, such matters as nondestruction, conservation, maintenance, restoration, and so on.

Now substitute *supporting an environmental cause* for *protecting a natural environment.* Then an agent is *green* in this expanded laxer sense if the agent believes in supporting an environmental cause and takes some relevant action in that direction. Such an action component is essential: *green* does imply some *practical,* if utterly token, involvement.

The lax account connects in the right ways with testable sociological criteria. For instance, believing in supporting a cause can be cashed out in terms of joining an organization working for that cause, or even, reducing the slight action component still further, considering joining such an organization.[19] The appellation *green* applies not only to individual agents, but also to groups, coalitions, institutions, even corporations and governments. As a green government can meet similar lax conditions, it is evident that there really are very few requirements, even weak sponsorship may serve.

Many there are now trying to take advantage of green movements – advertisers, vendors, politicians, professors, merchants, even generals (with 'defence forces' retooling as environmental defence forces). Cashing in on the movement by producing goods or services presented as environmentally friendly does not thereby make parties doing so green, especially if what is supplied *adds* to overall environmental impacts (as do 'environmental markets' delivering environmental junk, sometimes portrayed as 'green goods'). The requirement of serving environmental causes has been lost.

The excessive generosity of the account arrived at, which allows agents with only passing or slight active beliefs in some cause to count as green, does not matter. For firstly, it corresponds moderately well to current lax usage, which allows almost any sort of *concern for* environments, coupled with an action component as slight as paying a token subscription, to serve. Secondly, it still exercises quite enough exclusive power. Many agents are not green. For instance all those tuned into possessive individualism, mainstream economics, or similar creeds are not green (i.e. not even lax green). Thirdly, needed discriminations can be made *within* the class of (lax) green. An important subclass of green comprises *ecological* green, where the causes include maintenance and protection of natural ecosystems (including wilderness) and these causes enjoy some primacy. 'Ecological green' is by no means a tautology.

To be *green* in more than a token fashion is to have some commitment to containing or reducing the environmental impact of humans on the Earth or regions of it. By virtue of the environmental impact equation (for a given region) that means commitment in the immediate future term to either

- human population reduction, or

- less impacting lifestyles for many humans , or

- improvements in technology to reduce overall impact.

Most humans in a region can do little to implement the third requirement, except to hope. Accordingly they cannot satisfactorily demonstrate their commitment in this way. A *genuine* green will meet both the other conditions. But since as big coalition as feasible is sought for a green coalition, marginal greens who meet only one of the requirements, perhaps in weak form, will not usually be excluded (from almost any green church). Issues of population afford a simple test (a necessary condition) for a genuine green in Australia; for that agent fails who supposes that while efforts (including the agent's contribution) to effect human population reduction should be invested *elsewhere*, as in parts of Asia or Africa, Australia can keep on growing.

From filling out the conditions delimiting green commitments, a *green platform* can be derived. Specifically it will include, along with the action clause, derivative directives as regards curtailing population, refining and limiting consumption, adjusting technologies, and consequent thereupon altering administrative, political and economic arrangements and institutions accordingly. What so results is a proper subplatform of the Deep Ecology platform (presented in Chapter 4 below). A fair approximation to the full Deep Ecology platform can be derived by adjoining requirements for *depth* to those for genuine greenness of the action platform; so result themes concerning intrinsic value in nature, and commitments deriving therefrom. In short, the well-known Deep Ecology platform accounts (as a fair approximation) to a *deep green* platform – which can accordingly supplant it.

As genuine green and green platform can be characterized given the initial notion of green, that is lax green, so similarly can other green compounds and modified forms be explained (in a preliminary way). For example, a *green ideosystem* is an ideosystem (an ideology, in the unbiased sense) that coheres with green commitments, and typically organizes and guides them. As genuinely green, it is a system that includes a green platform, and it could be defined in that way, as an ideo-system coherently extending a green platform. Similarly for green ethics, and for green philosophy (and also for other green notions that could be defined but are not presently needed, e.g. green goods, green consumers, green parties, green ideas and so forth). A green ethical system is a green ideosystem which is an ethic. Green ethics comprises such green ethical systems and their theory. Green ethics thus comprise a decidedly restricted subset of environmental ethics. For an environmental ethic in the broad usage, an ethic that addresses environmental issues, among others, may well contain themes that could not be adopted by a green agent, that may even operate against environmental causes. In consequence, a green ethics is an ethic that could be coherently adopted by a green agent, that does not run counter to environmental causes.

What we are here encountering is one of the ambiguities and elasticities in the notion of environmental ethic. An ethic which takes specific account of environmental matters need not pronounce in favour of these matters; it may be hostile to environmental causes. Commonly these days an environmental ethic will take environmental causes (perhaps correspondingly delimited) as fitting within a setting of human concerns or interests, or more generously sentient creature interests and aspirations, but not independent of or exceeding features of that base class. But there are certainly prominent ethics, accounted environmental, according to which animals do not have interests or rights (so insofar as they are treated with decency and respect it is on some instrumental grounds or other); similarly for other ethics which regard the whole of nature as a human resource. These are not genuinely green ethics. Another ambiguity in the notion of environmental ethic, crossing over that from pro- through neutral to anti-environmental (i.e. broadly against environmental causes), runs between these bounds: between a narrow ethic concerned only or primarily with environmental matters (a restriction of a wider ethics to that domain for example), and a wide ethics which, among many other things, takes specific account of environmental matters. While we do not exclude a narrower use, we favour wider use. For one reason a narrower use is artificial; parts of ethics do not stand in splendid isolation.

From green to greening is an easy transition. Greening signifies, as usual, *becoming green* (it is another process notion). Greening of ethics has tended to lag behind environmental activism; seldom has it even kept up with, rarely has it anticipated, growing public awareness of escalating environmental problems. Such greening as has occurred has taken various forms – greening of ethics is definitely plural, greenings – but with three main types discernible:

- Green 'application' of standard ethics. Since, as already observed, mere applications may yield far from green outcomes, not all applications to environmental matters should count as green. (That is, to put it paradoxically, environmental ethics may well not be green! Less paradoxically, some of what regularly pass as 'environmental ethics' are not green.) Only when there is at least reasonable prospect that green results will be delivered, should such processes be considered as greenings;

- Adaptation or extension of standard ethics to accommodate environmental causes, for instance to help resolve environmental problems environmentally. Adaptation of utilitarianism to animal liberation ends (by appropriate enlargement of the base class of utility holders) affords a prime example.

25

- Development of new nonstandard ethics, superseding established ethics, in order to further environmental causes.

As will appear, the three main types correspond, more or less, to the divisions to be made into shallow, intermediate and deep.

One of the controversial issues of the 1970s – whether a new environmental ethics, a genuinely green ethics, is needed – raised the question of the adequacy of the first type of approach, as opposed to the second and third approaches (which tend to merge into one another). The third type of greening, transformation to new green ethics – not a mere painting green, or green face-lift, or green extension of older standard ethics – will eventually be pursued, and promoted, in what later follows. It will be pursued not least because it turns out that, environmentally at least, *all established ethics are inadequate.* As to why, reasons will be assembled in what follows too. But main reasons are, in brief, these: that each established ethics fails in at least one, and normally in several, of the following ways (summarized using technicalese):

- the class of agents is illegitimately restricted to certain humans. Simultaneously, usually, the base class in terms of which ethical evaluations are made is illicitly narrowed, to the same privileged class of humans;

- components of the full action scenario (that diagrammed in Figure 1.1) are improperly amputated (variously, motives, or consequences, etc.);

- critical elements in the ethical nexus are reduced, without warrant. (All of subjectivisms, emotivisms, and naturalisms get dismissed on this sort of score.);

- unsatisfactory theories are advanced of some main kinds of moral terms and judgements: deontic, axiological, and so on. (For instance, it is claimed that what is right is what maximizes present utility, or what tends to maximize social stability, or biodiversity, or...)

- uninformative or implausible theories are proposed as to how ethical information is acquired. (Thus, for instance, intuitionism and its variants, moral sense theories, and so on.) All too often, information acquisition is also conflated with meaning.

With few exceptions, accounts that have been offered hitherto of the greening of ethics are defective. They are defective for several intertwined reasons. Firstly, they do not exclude ethical exercises which, while addressing green issues, do not further them, but may effectively reiterate and help entrench prevailing prejudices. For example, mere accumulation of essays in what passes as environmental ethics is taken to signal greening, when many, perhaps most, of those essays are not green. Secondly, they do not include the interestingly new types of greening, new-look and new ethics, but take greening to reduce to just a further application of standard ethics. Green ethics are simply more applied ethics – where applied ethics is the application of (standard) ethics "to specific issues or areas of practical concern".[20] Interesting green ethics, deeper forms especially, are not further applied ethics. The greening they propose means a radical change in the subject itself, ethics itself, not environmental applications of it. Thirdly (while correctly rejecting the second sort of account) they offer a misleading historical explanation of how greening and like radical changes occur, through expansions of ethics.

The first account has already been dismissed. Reasons for dismissing the second demeaning account (which tends to render environmentalism in ethics as like activity of yet another practical pressure group pleading for special interests in a peripheral area) have been indicated, but they bear elaboration. Most work will be devoted however to dispelling the third account.

Before the 1970s there were but sporadic attempts at greener ethics, isolated proposals outside philosophy for a change in ethics and a few principles that might fit into such an earth ethics.[21] Since then things have changed, even professionally in philosophy, but not very much. For, alongside the greening, and obstructing attempts to change ethics, has stood a characteristically conservative professional reaction, aiming to assimilate environmental ethics within prevailing ethics.

Notice that when a new subject arises which threatens comfortable status quo arrangements, there are commonly, as with colonization, two stages that are gone through consecutively: extermination and assimilation. First, there are attempts to reject the new subject out of hand, to rubbish it, to show that it is incoherent, unnecessary, undesirable, or similar, and sometimes all of these things at once. Second, when these attempts fail, as they often do, what is being colonized or conquered refusing to disappear altogether, there are efforts at assimilation. The renegade new subject is brought within the compass of prevailing arrangements, it is shown to amount to nothing more than a branch or minor extension of what is already established and accepted, and so no serious threat after all. But rarely, then only after a considerable struggle, is such a new subject given satisfactory standing, the prevailing wisdom acknowledged as defective.

There are many examples. For a reasonably simple example, consider the tirade against modern modal logic which gathered considerable philosophical momentum about mid-century (a campaign centred upon, if not orchestrated from, Harvard University, an institution which, in an earlier creative phase, had contributed significantly to the development of modal logic). So, when the tirade failed to exterminate modal logic, a more liberal classical logical theory could easily pretend that modal logic, which included classical logic as part, was at most a special extension of classical logic by some curious additional elements (namely modal functors, such as necessarily and possibly). But there have been repeated efforts ever since to contain even the special extension entirely (for instance, by syntactical or functional reductions of modality).

Assimilation exercises propose to construe environmental ethics as a further *application* of ethics, of established establishment ethics, that is. Environmental ethics is a further application of ethics, alongside business ethics, sexual ethics, medical ethics, nursing ethics, and so on, applying allegedly tried and tested ethical principles and methods in the environmental area. What is missed by this assimilation is exactly what makes green ethics different. The adequacy of tried and tested principles, which application presupposes, is no longer granted; the old (anthropocentric) values are under question, and a structure assuming them cannot be applied without begging crucial issues. Green ethics are morally, and modally, different.

Further elaboration of these points returns us to an overarching theme, already broached: the inadequacy, given deeper green requirements, of all established ethics. In outline one of the arguments to this theme runs as follows: – all established ethics answer back, in one way or another, to humans or persons. While *not* covering merely interhuman or interpersonal relations, as some critics have alleged, the assessments made of matters that fall outside a human nexus are presumed to depend entirely upon features of humans, such as their interests, preferences, satisfaction, rationality or similar. But that, with ill-disposed people (like too many of those living) or in the absence of people, leaves environments destitute of the value they undoubtedly have according to deeper green ethics. Through due recognition of environmental value, in particular, ethics has begun to green, deeply green, in ways that preclude application recapture. How they have managed to green has also been misrepresented.

Apart from the application fiasco, a different, also a misleading, account of the greening of ethics now enjoys considerable vogue – what we call the *historical extension* thesis. According to this thesis, ethics was progressively extended, over historical time, to include ethically wider and wider domains. It was extended in the past from the family to the tribe to the region, then more recently to the nation-state, then within the nation to suppressed groups such as indigenous peoples and women, and without the nation to foreigners, and thence to all humans. Now we are in the process of extending ethics yet further – this is its greening – to encompass animals, perhaps other life forms, possibly the earth itself.[22] For all that the thesis can boast a fine pedigree and top supporters (from Schweitzer through Leopold to distinguished contemporaries), it is seriously astray. Similarly astray are analogous, though already less plausible, theses regarding greening of other subjects, such as philosophy and theology.

That ethics were not so limited in the past and did not develop by extension is revealed in several ways, for instance, reflection on features of ancient ethics (e.g. Biblical ethical attitudes to nature), confrontation with alternative sustainable cultures, and so on. To realize that something is amiss with such ethical extension, it is enough to inquire about the proper timetabling of this progressive enlargement of ethics. When did women enter into ethics? Have animals? If not when are they likely to be included? Women had roles and a place, typically an inferior station compared with free-men, not only under tribal and indigenous ethics, but also in ancient ethics, indeed virtually as far back as historical records on such matters reach. For example, their subjugated place was already established, but was reexamined, in classical Greek ethics. Similarly in modern ethics, women certainly figured, whence in part the rather different nineteenth century discussion of the subjection of women, long before the US Nineteenth Amendment of 1920 granted certain *legal* rights to *American* women, rights already achieved by Antipodean women.[23] Analogous damaging observations hold also for animals, and to a lesser extent for trees, forests and other conspicuous parts of natural environments. Bibles, and like ancient authoritative texts, prime sources for much contemporary ethical wisdom, have messages on all these topics, messages still heeded (unlike their scientific pronouncements). So do more extensive works on ethics. An early modern work, Spinoza's *Ethics*, to take one instance, includes an infamous passage on the proper treatment of animals (while such a hard doctrine was conventional rationalistic wisdom, it does not sit so prettily with Spinoza's pantheism). Before the advent of the modern mechanistic ideology, under which animals were sometimes downgraded to mere machines, animals and trees were often in fact assigned rather more extensive moral and legal roles, and sometimes responsibilities for damage (witness their standing in medieval trials, recently rediscovered) – one feature, among others, that (however unfounded the practices may be) casts some doubt on the progressive extension picture. In many societies, too, women, slaves and domestic animals have long been parts of communities (literally also, since together within the same walls as more exalted ethical beings).

In Western traditions the characteristic spirit of community has not been limited to humans or even to certain favoured groups of humans. Albert Schweitzer shot very wide of the historical mark in his famous comment more than half a century ago, "The great fault of all ethics hitherto has been that they believed themselves to have to deal only with the relationships of man to man".[24] While certain humans, persons, not infants, not just men, were undoubtedly the privileged players, for instance in Kantian ethics, non-humans, and their relations to humans, were certainly dealt with in major ethical systems.

No doubt things have changed from period to period in ethics, sometimes for the worse, as with 20th century European interludes of racism and of greed. What expanded, however, when expansion of some sort there was, was not the scope of ethics, nor even usually the circle of agents, but what was given improved or *decent* treatment – was, as perhaps seen from outside, treated more fairly, assigned further rights, and so forth.

Ethics was never such, or so characterized, that it fluctuated with some favoured circle of agents, that for instance it applied only to certain upper eschalon classes of agents, such as those divinely gifted or propertied or free-men or whatever. But that did not imply that items outside more favoured classes were not considered, and, while differentiated, ascribed stations and prescribed. Ethics comprehended all these different classes. For example, although infants typically fell outside favoured circles, they obtained differential treatment, often according to where they belonged (your family, or some remote tribe), and were treated ethically differently again from animals and from forests. Generally all of these items were comprehended ethically, and their treatment was regulated indirectly by prescriptions and principles applying to ethical agents. An alternative picture to extension thus emerges: that of differential circles within the scope of ethics. Most historical ethical systems were so class based, wrongly class based no doubt, because the classes were not based on relevant ethical features.[25]

Further troubles with extension notions are nicely illustrated through Leopold's exposition, an approved model widely copied since. Leopold's essay "The Land Ethic" "begins with the story of Odysseus and the slave-girls, lifted with only minor changes from the 1933 paper and [with] the concept of ethical evolution. [Odysseus] returned to his Greek homeland to hang, on a single rope, a dozen slave-girls accused of misbehavior in his absence. ... The point ..., Leopold explained [given that] Odysseus was an ethical man, ... was that slaves were property, and as such outside Odysseus's ethical community. ... With the passage of time, ... an 'extension of ethics' occurred. Slaves became people ..."[26] However a more satisfactory account than the extension one can be given. Namely that the slaves, who were always people, ceased to be property also. Of course they were never property in quite the way a necklace of stones may be; the slave-girls were accused, perhaps correctly, of *misconduct*. They were *already* within ethical scope, a feature that Leopold tries to evade by presenting "relations with [slaves as] strictly utilitarian", surely however part of standard ethics. The complementary

31

strategy consists in restricting *community* – an obvious community of humans comprises Odysseus's household, women, children, slaves, and so on – to an *ethical* community, not characterized, but some more privileged circle. On the more satisfactory account what occurred was not an extension of ethics, but a removal of unwarranted differentiation between overprivileged and underprivileged people. No doubt a change in ethics occurred, the permissibility principle already ethically relating masters and slaves, to the effect that it is permissible to hang slaves without due process, was overturned.

Historical extension themes, and associated expansionary ethical evolution, look more plausible if represented for *communities* or differently, for *rights*. Observe that so restated much of ethics is no longer covered. For an ethics may venture nothing about communities or rights (relevant terms may not occur among primitives). For example, utilitarianisms include no propositions concerning communities (no doubt in an expanded theory some such statements could be adjoined); pure act utilitarianisms correspondingly say nothing about rights. As it happens, the notion of rights has risen to prominence only in recent times; it seldom or never figured in ancient or in indigenous ethics.

By contrast, the notion of community did enter into ancient discussion, but in ways that do not assist extension themes. For communities were definitely plural; different cultural communities were widely recognized, although not necessarily in ethically relevant respects. Such essential linkage was with the cognate *ethos*. Both 'ethics' and 'ethos' derive from the Greek 'ethos' meaning custom, character, people or system. According to the Oxford English Dictionary 'ethos' has come to mean in English "the characteristic spirit of a community, people, or system", thus playing a cultural rather than ethical role. The ethos of a business is very different from, and only loosely coupled to, its ethics. Cultural communities and ethics have drifted in different directions, particularly with the advent of ecology which placed a new emphasis on communities, which were not of course confined to agents. Trying to effect repairs to extension themes through restriction to ethical or moral communities will not succeed, because whatever these are – do they include immoral humans, outcasts, or differently infants, hominids, altruistic dolphins – they will include differentiated subcommunities – petitioners, responsibility-bearers, and other ethical classes. There is no single moral community. *The* moral community (which is alleged to expand over historic time) is a misplaced definite; an extension theme based on 'it' is based on an error.

What about rights? Since the notion of rights got into wide circulation, surely the class of right-holders has been progressively extended, and differently right-holders have acquired more rights?[27] The latter is certainly not the case; recently smokers have been losing rights (both legally and socially) as regards where they smoke, and many more serious incursions on rights have occurred. Nor is the situation so clearcut as regards right-holders. Second-class citizens, prisoners of war, members of inferior castes, and so on, are all right-holders; even slaves normally held the equivalent of rights, they were entitled to certain basics (even if unscrupulous masters found it easy to evade responsibilities). So it is simply not true that slaves or women suddenly become right-holders, as ethics expanded. Rather they were granted rights, certain freedoms for instance, they had not previously had. It could even be argued – a conservative theory could take this line – that the class of right-holders has not changed. However we consider that the class of de facto rights-holders is changing. Most obviously, animals are now widely taken to hold rights, whereas they would not have been taken to do so in the 18th century. (The conservative would say that they held rights then, though the matter was not recognized: whence 'de facto'.) Such shifts do not help ethical extension; for animals are not thereby brought within the charmed ethical circle, as they were a subject of ethical assessment all along.

A real greening of ethics involves, in the first instance, a redistribution of values in way that revalues upwards various environmental items. In more flamboyant terms, it involves an environmental transvaluation of values. This transvaluation can occur either directly, the deeper ways or indirectly, the shallower way. Under a *direct* transvaluation, items which were previously assigned no intrinsic value (and perhaps little or no instrumental value) are now taken to have (always to have had) value in themselves. There is a very wide range of items that are candidates for such an overhaul of values, ranging from individual living creatures to whole ecosystems. The overhaul may also include a devaluation of items that were previously assigned excessively high values, such as certain sorts of humans. Under an *indirect* transvaluation

33

an analogous redistribution occurs as regards that feature (or cluster of features) that is considered to carry or reflect value, such as sentience, preference-having, self-autonomy, interests-possessing, or whatever. For example, not only humans but other individual creatures, and perhaps whole communities of creatures are taken to have interests, and so indirectly, through the linking with value, to exhibit value.

Naturally a greening of ethics includes much more than a trans-valuation of values. That transvaluation includes a wave effect, rippling almost everywhere, through the rest of ethics. For example, there is an impact upon basic deontic notions, because certain sorts of treatment of environmental items is consequently forbidden or required. Given, for instance, that degradation or destruction of intrinsically valuable items is normally prohibited, various sorts of environmental practices are accordingly prohibited. What is impossible, what is restricted, what is vandalism, and so on, changes in scope. Likewise justice changes from what it was or had been supposed to cover, because there are items, previously neglected in moral affairs, to be given their due.

No doubt the greening of ethics can be described in other ways than through a transvaluation of values and its impact on the remainder of ethics. For there are other ethical starting points, other ways into greener ethics. One is from deep moral attitudes, drawn out (or awak-ened, even 'remembered' in Socrates' terms) by decisive examples, such as that of the Last Person. The last person in the world, who therefore answers to no other person and interferes with no others, proceeds to significantly change the environment, for example wantonly or other-wise destroying valuable parts of it.[28] The immediate reactions tends to be in terms of *wrongness* of what is done, occasionally in terms of vandalism or the like (thus a redescription may be attempted in terms of character blemishes). But these judgements (of deontic or virtue ethics form) normally imply values. For example, that the Last Person's destruction is wrong is not independent of the value of some of what is destroyed.

What has been told of environmental ethics, its elasticity, of the divergence of green ethics, and of the greening of ethics can be retold for many other subjects, for economics and politics, and for other parts of philosophy than ethics, such as metaphysics and philosophy itself. The term *ecophilosophy*, which has gained some circulation, covers roughly what we should account green philosophy.

Many and varied are the aims of environmental philosophy ethics, from the lofty to the trivial, from delineating a colourless philosophy of nature, or from showing the transpersonal Way, through showing there is no cause for concern or action, to analytic wrestling with recalcitrant concepts in the traditional theory of Yalvard environmental dance. All traditional divisions of philosophy and its broad periphery can be rendered environmental, by being duly switched onto environmental matters and problems, and some such as old-fashioned nature-philosophy always were. Much of this sort of environmental philosophy is not green, but some of it admits of greening. Part of the intent of more ambitious greener environmental philosophy is frankly salvational: to lay foundations of hope and practice for saving the non-human world, perhaps providing therewith arguments for the importance of the bulk of the world that goes beyond immediate trivial concerns. These reasons are mostly of a metaphysical or an ethical nature, rather than, say, an economic nature (but undercut those of a narrowly economic character).

Although our focus will be upon green ethics (not applied ethics, for which we obviously do not have much time), parts of environmental philosophy are effectively unavoidable, because ethics is informed by and responds to other reaches of philosophy, logic and metaphysics particularly. Because of these interconnections, the idea of rendering ethics autonomous, an ethics without philosophy scheme, is unviable.[29] As well there are converse schemes in circulation which would dispose of some of ethics (typically the whole deontic sweep) or even all of ethics in favour of other parts of philosophy (typically metaphysics or philosophy of action or of evolution) or altogether. Those schemes are of remarkably diverse character: some are variously based upon metaphysical or historical determinism, some on the supposedly intractable features of ethical judgements (e.g. their verificational meaninglessness, unintelligibility, incoherence, indefensibility) some on the distasteful or damaging nature of certain classes of ethical judgements. Some of the dreary claims involved, while outside our main focus, will be eased aside in passing; others we do consider explicitly later since, within Australian philosophy, they have been directed against green ethics.[30]

Chapter 1

ENVIRONMENTAL PROBLEMS: THEIR CHARACTER AND CLASSIFICATION

Action and concern are directed above all at environmental problems, at addressing and resolving these problems. No problems, no cause for concern, no need for action. So it is particularly important to gain some grasp of the range and character of these problems. It will come as a surprise only to those not versed in academic ways that, despite the new extensive literature, there is no very satisfactory well-organized classification of environmental problems.[31] Mostly what is offered comprises incomplete and disorganized lists. A well-organized classification offers immediate benefits, such as: revealing something of the greater spread of what environmental philosophy should be addressing, by contrast with what it does so far treat better or worse, and mostly worse; and further enabling dismissal of most proposals for dealing with environmental problems as inadequate, for the reasons that they address only certain problems or parts of the problems; and so on.

An *environmental problem* is analytically, a problem of environmental cast, a problem concerning some relevant environment. Obviously, then, there are two components to explaining an environmental problem: giving some account of *environment* and, more demanding, some account of a *problem*. Much that is relevant to what follows can be extracted from these mundane-looking exercises.

The critical term *environment* means: surrounding, (surrounding) neighbourhood, region, including, therewithin objects, habitats and circumstances of the given surroundings. Breaking the term down into its etymological operational components yields a similar result: *environ-ment* is then the output of result, of (courtesy of the original French) circuiting around.[32] Plainly the term is implicitly relative: it is always the environment *of*, of some location. It is the surrounding region or district of that location, its neighbourhood, and what it comprises. That is an environment as normally conceived, namely working outwards from a given location; working inwards it is a restricted region of a world. As in mathematics, it is a neighbourhood (as duly axiomatizied in topology); and it has the general properties of such mathematical items. That is the techno-logical essence of the notion. Normally, within these confines, it is a geographical neighbourhood, in a familiar extended sense of 'geographical'. Like such items it is very general (though as a region of a world quite concrete enough), and unrestricted (except geographically) as to kind: so natural environments, built and urban environments, internal and external environments, biological environments, and solar and deep-space environments, are all proper types of environments.

It is an almost immediate consequence that ecological systems and ecologies are only some among environments. That is, the extension of 'ecology' is a quite proper subclass of that of 'environment'. Of course, the consequence turns also on the meaning of 'ecology', and its derivatives and pluralization, which however is (for once) exactly what the dictionaries tend to present; namely a subject, "a branch of biology", "dealing with living organisms' habits, modes of life, and relations to their surroundings".[33] *An* ecology is accordingly an environment of living items (one or a set, normally a community) where the modes of life of these items unfold, so that what is dealt with in ecology occurs there. It is an environment, in dynamic aspect, of living items. Whence the proper subclass results; environments devoid of living creature are not ecologies. Environments of planets devoid of life are not strictly ecologies. Rocks and mountains of such an environment are not alive; they are not strictly dead either; that distinction does not apply (of course a volcano that has ceased erupting is, metaphorically, dead). Organic and like images are likewise misplaced.

A further corollary, of some importance subsequently, is that environmental positions predicated on features of ecology (such as ecologisms like Deep Ecology, transpersonal ecology, shallow ecology, social ecology) are prima facie inadequate for comprehensive environmental purposes. For what happens to non-ecological environments, environments without identifiable life forms, may *matter*. These matter, ethically and otherwise, as when Japanese entrepreneurs actively plan to construct grand cemeteries and mausoleums on lifeless luckless planetary satellites, or when American scientists decide to terraform planets or to muck with splendid solar formations. Deep Ecology as formulated, for instance, cannot accommodate such problem cases – without implausible extensions or reduction stratagems, precisely the sorts of devices ingrained in shallower positions it was supposed to avoid. It also follows that there is nothing in the notions of ecology and environment that justifies the denigrating contrast of environmentalism with ecologism that has worked its way into a segment of environmental literature.[34] Entirely without basis is a scope-narrowing redefinition of environmentalism which could render it managerial, technocratic or engineering environmentalism. To the contrary, technocratic environmentalism is one type of environmentalism, ecologism is another type, environmentalism itself being some systematic theory and principles concerning environments.

For all we have said in favour of the term 'environment', it too has its limitations. For one thing, it regularly represents the other: the background, periphery, margin or shadow *of* something else, which is what gains prime exposure, the background relative to what is featured. But what we seek is something that is itself featured; for example, something that is itself spotlighted while humans play out their humdrum affairs in the dim background. A connected limitation derives from shallow deflation. What is environmental gets treated as a resource, a resource of, something for the use of, what it is considered environment of. Such a widespread deployment is evident in such combinations as 'environmental economics', which is a topical updating of 'resource economics'. Of the three sources of economic production – land, labour and capital – 'land' was first expanded to resources, and subsequently, with newer fashions, was expanded again to environment. Under the last expansion it includes not only sources, but sinks as well, economic products considered as externalities to production processes, such as pollution and waste. The environment, so deflated, functions again as merely instrumental background.

A disappointing upshot, *if* these regular associations are taken seriously, is that there are no really satisfactory terms, no untarnished expressions, for the philosophical and other purposes we hope to advance. Either we invent something new, always a risky practice, or we live with what is available or some adaptation thereof. Here we take the latter courses. We shall use 'environment' (and its compounds) *without* the background-only and resource-use associations (technically we use the geographical restriction of a topological characterization, which duly elides such associations). We shall also deploy 'green' as a quasi-technical term (to be read as 'gre-een') in the same re-generated sense.

Problem is even more of a problem. Dictionaries and other relevant literature offer less guidance, despite the ubiquity of problems, their spread everywhere (with similar terminology even in all commercial languages), and their expanding discussion. In brief, however, a solution looks like this: a *problem* is a set-up, where accordingly arrangements and conditions are thus and so, where there is pressure to find an acceptable procedure with such and such outcome. Much in this compressed account calls for explication. Set-up has that connotation it has acquired in logic; for most purposes *situation* could substitute, for some purposes, state-of-affairs or even proposition would serve. The pressure involved can be of various kinds, ranging from a challenge, as with a chess problem or other games, to a requirement or obligation of some sort (e.g., 'something must be done'). Here is one of two main points where broadly normative character enters into the matter of problems; what counts as a problem is far from a purely factual issue. This first point is of deontic cast; the other point, concerning acceptability, is of evaluative cast. The latter point is simply that not any kind of procedure will fulfil presumed requirements; what is acceptable has to measure up to certain standards of adequacy. A procedure is an agent-involving process, where a process in turn is a time-directed function from input to outcome states.[35] An acceptable agent-interacting process with a felicitous outcome is of course a solution. There is no guarantee however of a solution. Not all problems, not all environmental problems, have solutions.[36]

The way in which normative considerations enter into the very notion of problems, and correspondingly into spanning solutions, sets the scene for a cross-classification of environmental problems that will be important in what follows: into shallow and deeper problems and corresponding solutions. They can differ, considerably, as critical indicator problems show. For example, while there is a deep problem as regards continuing loss of 'minor' species of creatures, there is no corresponding shallow problem. For as extermination of 'minor' species does not matter shallowly, there is no *pressure* to find an *acceptable* procedure to slow losses down ('minor' implies anything independent of known human interests).

As to adequacy of the account of *problem*, virtually all the other things said about problems, and other cross-classifications made of them, fall out of, or else can be coaxed out of, the account elaborated. For example, the difficult/easy classification applies directly to ease of finding an acceptable procedure.

Environmental problems are not simply *in* environments, as for instance within agents' environments; they are problems *concerning* environments. Many of the problems that agents have are not environmental problems, even if unlike mathematical and other more abstract problems, the problems involve items occurring in agents' environments. For example, to list a comical agent's problems in the usual condensed fashion (naturally the condensations admit of proper elaboration): the agent's sink is blocked, his car will not start, his wife has left him, etc., etc. These may be problems concerning items in the agent's environment, but they are not environmental problems, they do not concern surroundings *as such*. More testing examples (e.g., all the drains in the agent's neighbourhood are blocked) soon reveal that 'environmental problems' is a vague notion, beset with many borderline cases.

In order to reach an improved classification of *environmental* problems, and therewith of environmental issues, it pays to look immediately to the notion of *impact*. (This also offers a further fix upon environmental problems.) For it is through environmental impacts that environmental problems manifest themselves. Impact is the operational aspect, the basic process: 'impact', that is, in the broad (original) sense of *impingement* (making something do something). With such 'impacts', no collisions, no rapid strikings or the like need be involved; groundwater pollution, for instance, may take a slow undramatic unseen course, though it may impinge, insidiously, in perhaps life-threatening ways. Not all impacts are *negative*; some impacts can be positive, beneficial, such as reforestation of damaged lands. Only negative impacts of sufficient extent and intensity constitute *problems*.[37] It is these impacts that concern us.

Consider now *environmental impact in a dynamic setting*. An impact is a kind of process which operates over time *from* somewhere, *sources*, *into* something, *sinks*, parts of environments. In a process diagram, it looks like this (as plurally generalized):

40

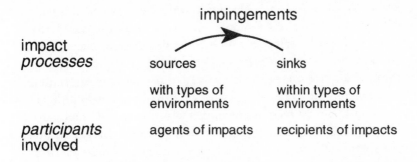

FIGURE 1.2: Environmental impact in a dynamic setting

Expanding upon the elements of this process diagram will deliver facets of a sought improved classification. Firstly, given independent background experience, especially with environmental impact assessments and equations, a further two-way classification is suggested: in terms of

- sorts of impacts, and

- types of environments.

The latter 'types' framework is wide enough (in obvious respects too wide for a tight classification) to include as well as broad geographical types both special concentrated sources such as mines, power stations, factories, and the like, and special sinks such as rubbish dumps, sewerage outflows, and the like. Initial sinks may not of course be ultimate ones. For instance, owing to flow systems on the Earth, much persistent industrial pollution ends up widely distributed at or above polar regions.

A working classification of types of terrestrial environments can be drawn from environmental geography. A standard classification will include savanna, forests (of various sorts, such as hardwood/softwood, tropical/temperate/boreal, etc.), agricultural lands, deserts (too many the products of past human activity), urban conglomerations, polar regions, atmospheric zones, and so on. Data on types of environment can often be represented in the usual way on maps, with special sources and sinks also depicted. While such standard classifications and representations still serve main purposes, no doubt improved classifications are available or can be devised. For instance, after the fashion of factor analysis, critical classificatory factors can be structured and applied, to pick out types. As things stand however, even standard classifications are underutilized. Many significant types of terrestrial environments obtain little or no attention in most surveys of environmental problems and issues, particularly as dished up in environmental philosophy. Indeed shallower environmental philosophy typically gives little or no attention to environments not heavily exploited by or of high significance to humans: ignored or downplayed are, for instance, remote wilderness, extraterrestrial environments and past (pre-human) environments.

But such a purely 'geographic' framework is not quite wide enough; it is important to factor into types of environments, expected inhabitants of those environments, which are the recipients of impacts (the impactees). Dynamically these will include not only present inhabitants of environments – animals, plants, native humans, for land-surface environments – but also expected future inhabitants, whose livelihoods or very existence may be impaired. There are two ways of incorporating these sorts of elements in the general scheme: either expanding environmental components, by having zones and habitats with kinds of expected inhabitants; or distinguishing participants in the impacts, as givers and receivers. Here the latter course is preferred; we distinguish also then

- kinds of recipients, and as well of course

- kinds of impactors.

Among the recipients of impact are many minority or underprivileged groups, the dominated or oppressed; for instance, animals and plants whose habitats are degraded or eliminated, or whose treatment in the interests of humans or of science fails to meet standards of decency and adequacy. While kinds of impactors are diverse, including such superagents and nonagents as Nature, Chance, God (as in 'acts of God'), and so forth, as well as animals, the main impactors of interest are, presently, humans and their industrial schemes. To a much lesser extent, feral animals and weeds are serious broad-area impactors.

Sorts of impact through human activity divide, economically, into two broad types, depending on their economic relation to goods and services: namely into

- production and

- consumption.

Each of these divisions decomposes further according to whether the activity concerned is that of an individual, corporation (still 'person') or group, whether it is small area and so (when it matters) concentrated or instead broad-acre. In principle these divisions yield at least 12 classes (i.e. 2 × 3 × 2, with more should we distinguish intermediate area operations), but some are relatively unimportant (e.g., isolated individual consumption, unless some consumption greats evolve, Mr Bigs of pollution, etc.), and many are similar in relevant aspects (e.g., a concentrated smelter that pollutes on intermediate area and a clear-cut forestry operation that degrades a similar area).

On the production side what is included – the factors of production – can be depicted through another process diagram, or flow chart:

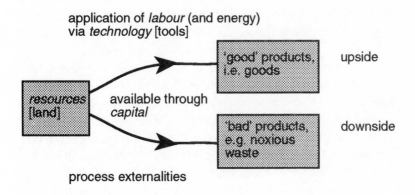

FIGURE 1.3: The factors of production

The diagram appropriately embroiders the stock oversimple economic story on the factors of production (with a stock list of factors *R, L, T, C* italicized), which however presents only the upside of production. The downside includes, part of what standard production stories leave out, both process and product 'externalities', of negative character.

The stock production relation depicted can be presented in functional form, with production *P* a function, *f* say, of the factors. That is, $P = f(R, L, T, C)$. That equation does not pretend to provide an operational procedure; it expresses no more than that there is a relationship of functional character, between the factors and the output P.[38] While there may be further factors that should be explicitly distinguished – *enterprise* is sometimes proposed (as to by whom, no guesses are given) – they are presumed not to upset the functional character of the relation.[39] It is however what is left out of this functional representation, that may upset its functionality, that matters especially environmentally.

Production includes many features, some of them thermodynamically inevitable, such as degrading of stock sources and delivery of pollution and waste, that have not been represented in the equation. The omitted features, extraneous components, though factors and products of production, resemble externalities; but the term 'externality' is normally reserved for certain features of market operations: omitted, ignored, unintended or like features. As the extraneous components of production not only resemble but succumb to rather similar classifications, to those of externalities of markets, let us adopt an obvious postmodern term to comprehend them: namely *extranalities*. Command economies with few or no market arrangements, and therefore few or no market externalities, as well as market economies, incur extranalities. Indeed the environmental extranalities of production systems of the command economies that held sway in Eastern Europe have become all too horrifyingly evident. No doubt *any* economic system, because bound to include economic production processes, will suffer extranalities. As a matter of definition, environmentally sound systems will reduce these extranalities to acceptably low levels (e.g. at the waste product end, though conversion of 'waste' material into needed products, cogenerative usage of waste heat, etc. – any safe remainder being well within absorptive capacities of relevant environments). Such sound systems represent an achievable ideal, from which all industrial economies are presently far removed.

While some notable extranalities count as individual (e.g., ill-health of workers) or as social (e.g., displacement of communities, social opportunity costs), several also rank as environmental, specifically all those significantly varying environmental value (and thereby varying environmental quality). A working *classification of environmental extranalities* can be gained by following through the elements and stages of production processes, which divide broadly into sites affected, processing, and products.

- SITES affected take in those of extraction, of processing, and of infrastructure, including therein storage, transport, and auxiliaries (which may, for instance, include whole townships for workers), and also affected surrounds. The *types* of sites now comprehend the whole range of terrestrial forms, under whatever geographical classification is adopted; for there is very little that falls outside the reach of projected mining and exploration adventures. The impacts, where negative, include again site disturbance and degradation, upsetting perhaps therewith both natural and historical features. Among the effects may be habitat degradation, species impoverishment, and so on.

- PROCESSING IMPACTS. To separate these more clearly from product impacts, take them as ephemeral, as opposed to persistent, with products considerably outlasting their processings. Among such processing impacts are noise, energy wastes (such as heat and radiation), congestion, and (ephemeral) pollution. Correlative to such impacts are capacities of environments to assimilate or absorb them, capacities sharply limited by biophysical features.

- PRODUCT IMPACTS. These include material wastes and persistent pollutants. But evidently product impact itself is much more extensive. Production processes may terminate, as far as some producers still assume, when the products leave the factory floor, farm gate or similar, but products like persistent waste live on, to face disposal or recycling when they decline also to waste. Full 'life cycles' of products matter environmentally. It is past time that these matters were satisfactorily assessed, and priced, and results informed consumer choices, improved choices.[40]

In assessing productive impacts, not only the production situation itself enters; so do alternative situations where other, perhaps more environmentally benign, processes operate. That is, improved economic theory takes into account a range of accessible alternative situations or worlds (which 'positive' economics has tried to knock down to environmental opportunity costs, or less). It is in terms of these that acceptability and satisfactoriness of productive modes and practices get assessed. Production normally answers to presumed consumption; without potential consumption most of it would be pointless.

On the broad consumption side, much is synthesized through a standard environmental impact equation, an equation usually intended to apply to humans in a given region. But it applies to any class of impactors in a region, through any kind of impact:

EIC. ENVIRONMENTAL $=$ POPULATION $\times$ RESOURCE $\times$ IMPACT
IMPACT   SIZE   USE   (through...)
(of...   (of...)   (of...) per   per unit of
through...)   member   resource use

The dots following 'of' are filled out by a description of impactors (strictly for significance, a population of impactors), and those dots following 'through' by a description of the type of impact concerned. Thus a fill-out following 'environmental impact' could be: of humans in USA through atmospheric pollution; or it could be: of possums in New Zealand kauri forests through defoliation, or: of hominoids in prehistoric Europe through overhunting. Creatures other than humans can induce severe environmental impacts.

The equation EIC is exact and analytic. It is just one, but a very helpful, breakdown of impact into components, which duly multiplied together return the total impact. With a little convenient adjustment

46

such an impact breakdown leads directly to three categories of environmentally relevant factors: namely in mnemonic form

- Population

- Consumption

- Technology

Thus in brief functional symbolization the equation runs:

$$EI \ = \ P \ \times \ C \ \times \ T$$

While the population factor corresponds exactly to the original, the consumption and technology factors afford only approximate representations. Consumption, however, could be made exact by replacing 'resource use' in EIC by 'consumptive use', i.e. in effect making the second factor consumption per capita (as 'resource use' is a dummy parameter it could be replaced by *anything* suitable, 'utiles' for instance). The third factor of impact per unit of resource use or consumption, does depend on technological efficiency achieved, and normally conversely, and accordingly is not too misleadingly summarised in the rubric technology. Even so, technology *applied* would be better, for instance because available technology may not be applied, or applied well, to reduce impact.

Before examining the important roles of this impact equation in organizing classification and discussion of environmental problems, it is worth investigating certain elaborations and variations of it. First, regional devolution. A more sophisticated environmental impact equation would reduce the aggregation in the equation by introducing distributional features, which indicated where and when impact really mattered. Suppose, for example, there are sufficiently independent regions (as presumed in the following regionalized impact equation, summing impacts of different regions 1 to n:

$$EIR. \ EI \ = \ \sum_{i=1}^{n} \ P_i \ \times \ C_i \ \times \ T_i$$

Suppose, for instance, we are investigating the terrestrial environmental impact of present humans through industrialization, i.e. EI (present humans through industrialization). Then, on an obvious regional breakdown, the impact is overwhelmingly dominated by three geographically separate regions: North America, Europe, and North Asia (primarily Japan). If it were not that the rest of the world were locked into this system (and its humans encouraged to applaud and emulate it), supplying much of the raw materials, taking too much of the expensive products and waste, and suffering the pollution effects that spill outside the offending regions, main problems of industrialization could be isolated in the high latitude Northern hemisphere. Specifically, the terrestrial environmental impact of humans through industrialization is the sum

$$\sum_{i=1}^{4} P_i \times C_i \times T_i$$

where 1 = North America, 2 = North Asia, 3 = Europe (West and East) and, what comparatively is negligible, 4 = the residue, the rest of terrestrial industrial regions. A walling off of these regions of Mordor would importantly confine most environmental problems.

Another relevant breakdown, which can be applied either to the initial aggregated impact equation or to its regional devolutions, consists in further analyses of factors. For example, the consumption factor can be split into consumption of non-renewables and consumption of renewables (or further down again to take account of recycling potentialities). Evidently consumption of some goods has a much heavier impact than others, so analysis is relevant to policy and prescriptions for reduced impact. For sustainability, for instance, consumption of non-renewables generally constitutes a larger problem than consumption of renewables, which, all going well, can be replaced, not merely substituted for. Again, the technology factor can be decomposed, for instance though analysis into sectors of an economy: manufacturing, transport, agriculture, mining, etc.[41] Again as there are considerable differences in technical efficiencies and coupled relative impacts, such an analysis will matter for policy and prescriptions.

What is more, we have now reached the stage where we have encompassed virtually all the main factors included in 'limits to growth' investigations.[42] Pollution, for example, is one sort of environmental impact, an overall output from the equation. Naturally, a dynamic picture of what is happening to impacts such as pollution can be better achieved by relativizing the equation, time-wise (e.g. to $EI(t) = P(t) \times C(t) \times T(t)$). Then too such relevant features as population dynamics, effects over time of immigration in a region, consumption patterns and so on, can be taken into due account.

The factors of environmental impact normally act in concert. Despite well-publicized attempts to load major environmental problems onto just one component, evidently *all* are normally important. While now, with many regions surviving close to the edge, an increase in human numbers may be enough to increase environmental destruction materially even without an increase in consumption or a change in technology, it is more likely that, as in the past, increased consumption and higher impacting technologies will tend to accompany, and interact with, increased population growth. Population growth will thus generate increasing demands for goods like food grains, fishery products, wood, minerals, and water. To make matters worse, these will generally be obtained in unsustainable ways, often exploitative and destructive of the environment. More generally, environmental impact equations serve to reveal the illusoriness of widespread assumptions as to how those impacts that are taken seriously can be reduced. Not only is there the illusory prospect of reducing impact to 'acceptable levels' by variation of just one parameter – C_i is that invariably favoured by socially-oriented shallower environmentalists (but its effect is liable to be swamped by P_i); T_i, in the shape of wizard technology, is that advanced by economists and technocrats (but such technology is likely to have its own impact and thereby to add to environmental problems).

Of course, there was, until recently, the possibility of moving further afield to new regions, where impacts are lower. There are familiar frontier practices (and corresponding frontier ethics, operating in frontier regions) which can be reapplied to elements such as industrialization and its components, pollution and waste. Of course such frontier practices are impoverishing, and too often destructive of further environments and many of their inhabitants. But in any case, though continuing in interior Brazil and Australia, they are now seriously limited by biospheric limitations of the whole Earth. Now new frontiers beckon for off-Earth adventures and impoverishment; these too will be subject to regional impact equations.

Chapter 1

What emerges is that all main environmental problems arise from broadly economic activities of humans (the dominant theory of which, until recently, evaded or ignored the initial problems), from their production and consumption. Any adequate general environmental theory has to address these broadly economic issues in due detail. Many proposed theories fail (as, for instance, Deep Ecology which does not sufficiently address ways to ensure reduced impact).

RESULTING CLASSIFICATION OF THEORIES OF CONTEMPORARY ENVIRONMENTALISM

Many of the popular proposals for resolving major environmental problems can be seen as reached by selectively latching onto certain components of impact equations and assuming that these are primary or sole causes of the problems. Such theoretical practice offers an important example of what is deeply ingrained in human intellectual methodology, excessive selectivity, and skimping on critical components.[43] What both theoretical investigation and practical experience discloses is the substantially interdependent character of the impact factors, that these factors tend to operate not separately but in concert (for instance, all of P, C and T operate together in gross regional growth or GRP). Resolution of larger regional problems generally requires address of all main factors in impact equations. (For instance, redirection of GRP growth, in gross and rampant forms a major antagonist of environmental objectives, involves adjustment of all factors).

On the basis of these observations we can predict what can subsequently be readily confirmed, the inadequacy of several main theories advanced in contemporary environmentalism: namely[44]

- the POPULATION THEORY, or neo-Malthusianism (according to which the prime factor to adjust is P). Main environmental problems are primarily due to excessive populations and their dynamics. Environmental problems are essentially demographic problems, and resolved similarly, that is through stabilization or reduction of population numbers, reduction therewith of population growth rates, and so on.

- the AFFLUENCE THEORY (C). The most significant factor producing ecological stress, and consequent environmental problems, is the striking increase in individual affluence and consumption (esp. in post-War OECD nations). Environmental problems are essentially over-consumption problems, resolved through addressing affluence and returning to simpler, less consumptive life styles.[45]

- the TECHNOLOGY THEORY (T). The crucial factor in environmental degradation concerns introduction of new damaging technologies (non-disposable, non-degradable products; radioactivity, toxic chemicals; etc.) and displacement or suppression of less damaging and wasteful technologies. Environmental problems can be resolved through technological change.[46]

- a SOCIAL ECOLOGY THEORY (prime factors to adjust: T and C). Human population is not an important factor. Social factors, especially those permitting damaging technologies and excessive consumption, are the prime source of environmental problems, which are secondary to human social problems.

Continuing in this fashion, plainly eight standard combinations can be generated from the three factors P, C, T. Two combinations, for both of which cases could be defended, appear to go unrepresented, namely PC and PT (though limits-to-growth positions come close to PC, as technology, which is applauded, is only of passing relevance). Remaining combinations are represented:

- a MATERIAL GROWTH THEORY (PCT). All the factors of the equation matter. All require attention in curtailing major environmental problems, which mainly result from excessive material growth.

While we have already stated that we do consider all the factors do matter, while we consider that deeper environmentalism should press for significant adjustments in all, we do not imagine that is the whole story by any means. For one reason, there could be significant improvements as regards all factors, perhaps with comfortable human futures assured, but with heavy environmental costs. For instance, the whole Earth is given over to environmental management, even turned into perfect gardens and towns, undisturbed by wild animals and invading native plants species. For another, there are further elements beyond or behind those material factors that figure in one set of impact equations: elements to do with environmentally unsound practices, attitudes, and values.

- a SPIRITUAL DECADENCE or MALAISE POSITION (under which none of P,C,T need to be adjusted). Material factors are not critical, and play only a derivative role. When spiritual regeneration occurs, when humans again live by requisite scriptures, environmental problems, which like health problems derive from ill-living, will fade away. One difficulty with this splendidly elevated stance is that humans are also heavily corporeal; and material factors tied to this exponentiating corporeality and its gross consumptive practice are critical to escalating environmental problems.

Many theories (properly) acknowledge that there is more to accounting for environmental problems and showing ways out of them than mere altering of critical factors. In principle such theories – which permit of rough classification in terms of the further elements invoked (economic, ethical, etc.) – can be coupled with each of the eight combinations outlined. We merely glance at a couple of options, which have been advanced.

- • an ANTI-CAPITALISTIC refinement of the technology theory (T+). The main theme is that capitalism, through its imperatives of growth and profit maintenance, forces the general introduction of damaging technologies (those who do not immediately adopt profitable but damaging technologies are allegedly driven out of the market, and similar). In brief, the reason for what the technology theory correctly identifies is capitalism.

- • • an ETHICAL or EVALUATIVE elaboration of the growth theory (PCT+). In broad form the elaboration asserts that environmentally damaging practices are supported through and often encouraged by disrespectful and exploitative attitudes towards natural items, attitudes that derive from defective ideologies, which underwrite dominance and exploitation. What is wanted for a turnaround is not merely considerable adjustment of PCT factors, but reformed attitudes underwritten by a different ideology, most directly a new value theory or ethics.

Obviously this sort of theory can be filled out in various ways, depending upon the kind and strength of the supporting relation claimed (e.g., whether there are ineluctable 'roots') and the type of ideology unveiled (e.g., Christianity, Protestantism, capitalism). Our view happens to be that *many* ideologies could have served to support damaging practices, as most religious and certain irreligious ideologies now do. There was nothing inevitable about the particular roots that historically evolved.[47] Certain levels of chauvinism, that new ethics could counteract, were all that were required, and such levels and more most religions and other ideologies freely supplied.

REGIONAL ENVIRONMENTAL IMPACT AND ENVIRONMENTAL CRISES

As regards environmental impact, Australia is still widely considered, despite two hundred years of strenuous Western effort, a comparatively lucky continent. Australia remains in a fortunate position for preserving large portions of its environment in comparatively natural shape. This cannot be ascribed to careful environmental management or practice since European settlement. Australians have established new records for degradation of a continent, with very impressive figures for species elimination, forest removal, soil losses, and so on, in just two hundred years of extensive vandalism. Moreover, though lagging developments in the North, Australia appears to have learnt little from hard environmental experience elsewhere. In agriculture for example, old mistakes regarding both dry land and irrigated farming are being replicated. The main cities are beginning to exhibit heavy pollution, congestion and conurbation, their sewer systems are proving inadequate, and regularly pour untreated sewage into enclosed waterways and onto beaches. And so on, through a distressingly familiar list. There is little cause for environmental complacency; all is not well, ecologically and otherwise, in the

53

formerly lucky country. But, while Australia now faces more or less all the same sorts of environmental problems as the rest of the world (nuclear pollution may be less, ozone depletion more), there remain many environmental riches that lie within its wide region, such as the Great Barrier Reef and Southwest Tasmanian wilderness. These and much else can, all going well, be saved as areas of world heritage and global significance.

It could be claimed, not too inaccurately, that Australia has become a bellwether territory. There are more reasons for such a claim than what remains, much of it is fragile and vulnerable enough. There are also significant environmental movements, dedicated to trying to retain much of it. It is estimated that Australia has more members of environmental groups in relative terms than almost any comparable country. "Documentary film maker Sir David Attenborough argues that if conservation fails in Australia then all hope of convincing the rest of the world of its importance is dead".[48] Similar statements have emanated from other touring environmental stars. Australian environmental groups are attempting, among other things, to convince Australia, and the rest of the world, that the environment is important and worth saving. If they fail, terrestrial failure appears likely; if they succeed, at least for the time being (all environmental successes are temporary, such is the perilous state of affairs), then an important regional beachhead for progress on turning back 'progress' elsewhere is established.

Failure cannot realistically be excluded from assessment of prospects. Environmental conditions, both regional and global, appear to worsen sometimes to crisis stage. For all the signs of deterioration, however, there is much dispute as to *whether there is an environmental crisis.* The outcome of this dispute matters practically. For, to reason practically, if there is a crisis or an impending crisis, and something can be done about it, then it should be done; whereas if there is not, and no real prospect of one; then *nothing* need be done, business and industrial development can continue as usual. Crises scare many people, they activate most; they are marvellous motivators to action. By contrast, no crisis may mean no action, muddling on, in worsening circumstances, as before. While the dispute over a crisis, or an impending crisis, is in part terminological, an issue we shall tackle at once, at bottom it may seem irresolvable, as it reflects rival value systems and therewith rival operational paradigms.

First to necessary analytic preliminaries. Among available dictionaries, the *Oxford English Dictionary* account of *crisis* shines through as superior. It correctly distinguishes the relatively precise pathological sense (item 1) from the vaguer popular use (item 3) with which (as slightly refined) we are concerned.

1. *Pathol.* The point in the progress of a disease when an important development or change takes place which is decisive of recovery or death; the turning-point of a disease for better or worse;...

3. *transf.* and *fig.* A vitally important or decisive stage in the progress of anything; a turning-point; also, a state of affairs in which a decisive charge for better or worse is imminent; now applied *esp.* to times of difficulty, insecurity, and suspense in politics or commerce.

The omitted sense 2 is an obsolete astrological one (parasitic on 1). Sense 1 is essentially a special case of sense 3, which now predominates.

Salient features of any genuine crisis can be usefully diagrammed as follows:

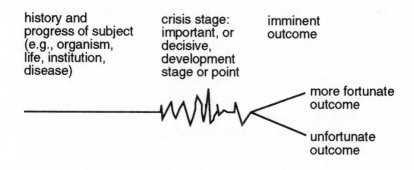

history and
progress of subject
(e.g., organism,
life, institution,
disease)

crisis stage:
important, or
decisive,
development
stage or point

imminent
outcome

more fortunate
outcome

unfortunate
outcome

FIGURE 1.4: Crisis process in given setting

Evident evaluative elements include elevation of the make-or-break stage itself, its ranking as important, decisive or even momentous in the progress of the subject concerned, and the rankings of the outcomes. For example, death or disablement are normally assumed to be rather bad outcomes. However an extraordinary reversion in values is what we encounter with those who are prepared to attest that gross biological impoverishment of the earth, for example, does not matter, that it is only human (monetary) richness that really counts. In such a setting an environmental crisis vanishes: we are not peering down a barrel at any undesirable outcome after all. Other lesser contextually supplied features of the characterisation portrayed (such as real possibilities of certain outcomes, immanence thereof, etc.) will appear as the dialectic proceeds. For we are already placed to apply the schematization, both against defective accounts and against the commercially convenient assumption that there is no environmental crisis occurring or impending.

One of the more remarkable cases against any environmental crises derives from Social Darwinism. Under it, life is essentially a struggle against others for survival. The destruction of other species and other businesses is just part of life, part of evolutionary and economic progress in which the weak and inefficient are inevitably mown down by the strong and efficient. Quite apart from the appalling values implicitly presupposed – that it does not matter what happens to others, only the interests of one's own narrow group matter[49] – a high redefinition of 'crisis' has been tacitly imported, under which destruction of entire peoples and degradation of entire continents do not constitute crises, indeed much horrendous change for the worse are not crises so long as they are elsewhere or other. But that illegitimately redefines 'crisis' to self-crisis, by adjoining a tacit location clause. Many normal crises will not be such self-crises, though some global crises may.

It is incorrect to say that 'a crisis is both an immensely dangerous period and one that will – if we do the right things – pass'. A crisis may have passed out of control; this is a commonplace situation medically, where it may be a matter of awaiting to see what eventuates (and hoping). It is accordingly wrong to go on 'It can be resolved by hard discussions now and by a preparedness to incur temporary hardships'. For example, *Beyond The Limits* itself (source of preceding quotes) outlines many crisis scenarios, some of which some of us can do something about – some with, some without hardship – and some where nothing can be done. It may very well issue in an unfavourable outcome, such as extinction of important species, death of relevant organisms, and so on.

Certain insinuations that there is no crisis tend to be based on a slide to the pathological usage of 'crisis', to signify a turning-point especially of a disease. That is neither the etymological sense, though transferred from it (Greek *krisis* meant decision or decision stage, but diseases do not make decisions in this way), nor the popular sense, where 'crisis' commonly signifies a period of difficulty or suspense.

For example, it is contended, assuming the pathological usage, that we may not face an environmental *crisis*, for ecological disasters and the accumulation of insults to planetary health are chronic, rather than critical, conditions.[50] Even in a medical setting these contentions are dubious; for many of the insults such as chlorofluorocarbon impact on ozone protection are very recent, not chronic (i.e., inveterate, lasting); and clinical crises may well be in process for certain crucial terrestrial systems, such as atmosphere and oceans. Outside such a medical setting the contentions fall to the ground.

There is a crisis. It is worsening. Not enough is being done about it. More contentiously, it will not be solved by persisting with shallow ways. Once the terminology is duly classified, it is not difficult to show that there is a crisis, several crises. So many critical indicators are bad and getting worse. Consider, for instance, present *daily* changes:

- loss of 116 square miles of rainforest;

- loss of 72 square miles to encroaching deserts;

- loss of perhaps 40-100 species;

- increase of human population by a quarter of a million;

- increase of chlorofluorocarbons in atmosphere by 2700 tons;

- addition of carbon to atmosphere by 15 million tons.

Evidently an environmental crisis stage has been reached. Moreover, the extrapolated and potential outcomes, enormous loss of rich habitats, of biodiversity, terrestrial greenhouse, and so on, are singularly unfortu-nate. Nor is it difficult to show that not enough is being done about it. There are, by way of illustration, no substantial programs in place to ensure continuing natural biodiversity. There are just a few programs trying to save fragments, isolated large species and the like.

Even assuming the medical analogy, with the patient now some nation or the human world, it can be persuasively argued that there is a crisis, both for nations such as the USA and for the human world. The investigations of works like *Beyond Oil* appear to show that the economic quality of American life is entering a crisis condition. Without some fairly drastic changes in regulatory structure, "a long-term slide in US per capita GNP will begin in the 1990's, reflecting declining stores of oil and gas", a decline that will impact on other major linked sectors of the US economy, especially agriculture.[51] What these studies accomplish, on the basis of systems theory computer modellings for the US economy, larger studies of the same sort adduce for macro-economic factors of the world. For example,

> if the present growth trends in [main parameters studied, namely] world population, industrialization, pollution, food production, and resource depletion continue unchanged, the limits of growth on this planet will be reached sometime within the next 100 years. The most probable result will be a sudden and uncontrollable decline in both population and industrial capacity.[52]

These sorts of claims of crisis and impending catastrophe have generated enormous controversy, especially from economic directions. While some of the economic commentary was simply scurrilous, some of the criticism had a point. Too much reliance was placed upon details of the systems modellings, which themselves incorporated too many question-able and variable assumptions. (As a result, they generated some now contradicted results, such as that tin supplies would be exhausted before now. A claim that tin supplies would probably run out if consumption trends continued, would not have been contra-indicated: projections are not predictions.) But the thrust of the background inductive arguments (upon which requisite detail was properly piled) was, and remains, clear: that there is a far from negligible probability of a sharp and painful decline in a critical socio-economic parameter in the foreseeable future if present broad trends continue.

The crises foretold under systems-analyses are, furthermore, only crises *within* given frameworks of assumptions; to illustrate in the case of *Beyond Oil*, that the American standard and style of living or something very like it, should persist, more sweepingly that the dominant American paradigm, which incorporates the American Dream, should not be swept away. But many social critics, not merely environmentalists, would say that it would not be such an unfortunate outcome if the American consumptive way of life were not maintained. In this different setting there is no clinical crisis; only something more like a return to reality from an impossible (long-term) dream. Even in the event of exceeding indeterminate limits to growth, though the outcome may be catastrophic, there is no serious threat to human survival or even to human comfort in some regions (e.g., to those sufficiently isolated or prepared); rather the outcome, which has been little investigated, may at worst somewhat resemble that from a non-nuclear world war, involving exterior removal of industrial capacity, possible famine, and population decline (but achieved without standard military practices, and not resembling nuclear winter). Whether there is a clinical crisis will depend upon how unfortunate the outcome is assumed to be; given a worst case scenario no doubt there would be a crisis by most standards, but because of the way it was allowed to happen. For a well-orchestrated decline in human population size and world industrial production would constitute no crisis, but would afford a cause for deep celebration.

The modellings developed in these systems studies are so far shallow in character, no parameters beyond those of standard resource economics being modelled or investigated, and they are substantially reformist in political upshot. No major transformation of social or political institutions and arrangements is proposed, only selectively minor adjustments. For example, almost all the familiar recommendations of *Beyond Oil* could be effected or substantially commenced in the USA tomorrow given political will (and suspension of vested interests); many are already in operation in other industrial countries, some in some US states. The recommendations are in outline these: increased fuel prices (no carbon taxes even are proposed); energy conservation, through efficiency, co-generation, public examples and incentives; investment in renewable fuels; promotion of low-input farming methods; and population growth control. Many of the recommendations of *Beyond The Limits* are similar[53] though much less specific (e.g., there is nothing about fuel pricing, or on gross military wastage). Overall they are more diverse, and more individualistic, possibly more far-reaching, but tending to schmaltz, which tends to increase as the discussion proceeds through "visionary [ideas], networking, truth-telling, learning, and loving". It is optimistically imagined that virtuous individual behaviour such as honesty, trust, truth-telling, optimism, and boosting information exchange (but few of them guaranteed under continuing business practice) will impact importantly on exponentiating resource parameters. There are *no* specific recommendations for structural change. While there is a heading "The next revolution: sustainability", neither any revolution nor sustainability are satisfactorily explained or appropriately enjoined.

Yet there are major environmental crises, composed, among other things, of many many regional crises. But nothing adequate is being done about most of them, though what to do is often superficially evident enough; still less are the overarching crises being satisfactorily addressed. To alter things for the better environmentally, there will have to be major changes both in attitudes and structure, ethics and politics. There is no sufficient sign that such changes are occurring on a requisite scale. For important reasons why too little is being done, often too late when it is, look to the sway of unfriendly ideologies, incorporating chauvinistic ethics. It is to ethics that we should turn to get to the bottom of the inaction business, to reasons for the lack of response to pressing problems, to crises.

PART I

Two Decades of Environmental Ethics: Types of Environmental Ethics

Since the early 1970's Australia has been a hotbed of environmental activity and, encouraged thereby, environmental philosophy. A main reason was that it became glaringly conspicuous in Australia, as else-where, that the environment was being degraded and devastated at an unprecedented rate. A moderately free, educated, and not entirely apathetic community could articulate its concern. Other reasons were also important, such as awakening appreciation of the appalling treat-ment metered out to many animals, and, very differently, those of ideological reaction: philosophical attempts to shore up and maintain the status quo. More recently different and newer reasons have been suggested. Perhaps because as both David Attenborough and David Bellamy have suggested, too flatteringly to Australians, if the struggle for the environment is lost here, then it is lost everywhere. Perhaps because as Nigel Austin, writer for *The Bulletin* suggests, "Put simply: mankind is destroying its home, the decline of which means a deterioration in living standards and ultimately the end of life. Man remains the only factor that can save the environment".[1] What Austin fails to point out is that the environment is the home of a multitude of species besides the prime problem factor, male humans, and that the deterioration in living standards applies to many species. Also the end of life is more than just the end of human life. If or when humans go, they will likely take many other species with them.

There are many reasons for *reassessing* humanity's relationship with the environment.[2] For one, Western civilizations have mostly related to the environment in terms of what they could extract from it or dump into it. The environment was a mere backdrop for what really mattered the affairs of humans and superhumans. It was infinitely resilient and totally suitable for nothing until touched by human hands. The result of this approach has been to erode and destroy the resource base upon which humans and all other life forms on the earth exist; to pollute the land, sea, and air, thus detracting from the quality of life; and to destabilize the ecosystems necessary to the health of the biosphere, thus jeopardizing human life. All of these reasons relate to the preservation of human life and the quality of human life. Many environmental philosophers would argue this is due, at least in part, to restricting ethics and what ethics are disposed to do to the relationships between people or persons. Insofar as other items, such as animals, plants, air, water, soil or the ecosphere were concerned, if they received any concern at all, it was in relation to the humans who owned them or were affected by their treatment.

However, as more people have become environment watchers, they are not only looking at what has been done to the environment, but asking why. Part of the reason for the mistreatment of the environment is that a genuine 'ethic of concern' did not apply to it. Of course an ethic of sorts did apply, but always one lacking environmental concern. Both the dominant and prevailing ethics could be summed up as, "It does not matter what one does to the environment as long as it does not adversely affect other humans". Various suggestions have been put forward for improvements, as to how to accommodate the environment ethically. The two most common suggestions are to polish up current ethics or ethical principles and seriously apply those to human/environment relationships, or to develop 'new ethics' that constitute qualitatively and quantitatively different ways of dealing with the environment. The first group sometimes argue, "what we need is not a new ethics but a new 'moral rearmament', a revival of moral dedication".[3] The substance of this argument is that our failure in reference to the environment is not a failure of our ethic, but our adherence to it, that is a failure to live by it. The second group argue, "The dominant Western view is simply inconsistent with an environmental ethic; for according to it nature is the dominion of man and he is free to deal with it as he pleases".[4] The substance of this argument is that while our current ethics may include principles that can be useful in developing an ethic that takes account of the non-human world, our current ethics are unresponsive to and do not take due account of environmental items and values and are opposed to a genuinely environmental ethic.

From these positions a spectrum of environmental ethics with varying shades of refinement can be developed. Environmental ethics in Australia and elsewhere can be divided into three levels: shallow, intermediate, and deep. But there is a contrasting position that is not an environmental ethic. The contrasting position is the short-term unrestrained, exploitative approach that takes little or no account of the environment ethically and typically assumes that the people can mould the land and the environment to their purposes and whims. The ruling passion of this position is often 'hip-pocket-racy'. Environmental consideration goes only as far as profit motive. If it is bad for business to be environmentally destructive, then some concessions are made for the sake of business and not the environment.

Shallow environmentalism is anthropocentric. Few constraints are imposed upon the treatment of the environment providing that treatment does not interfere with the interests of other humans. By contrast, however, with non-environmental ethics it does take a long-term view of environmental issues, and it does consider future human generations. For these reasons it is often described as resource management or husbandry. Intermediate positions can be distinguished as rejecting the notion that humans and human projects are the sole items of value; however serious human concerns always come first. These positions acknowledge the value-in-their-own-right of some at least of animals, ecosystems, forests, and other parts of the environment as well as the environment as a whole in addition to their value for human purposes. Deep positions are characterized by the rejection of the notion that humans and human projects are the sole items of value, and further by the rejection of the notion that humans and human projects are always more valuable than all other things in the world.

The whole core of environmental ethics, so approached, is setting environmental values. Not just setting a value on the environment, but determining how it is to be valued and where that value resides. For shallow positions it is the value for humans and resides in its relationship to humans. For deeper positions the environment is valuable in-itself and although there is debate about where that value resides, it does not reside solely in the environment's relationship to humans.

63

Shallow Environmental Ethics

We consider first SHALLOW ENVIRONMENTAL ETHICS, very much the predominant sort both outside and inside philosophy. Shallow positions remain anthropocentric; they do not move outside a human-centred framework which construes nature and the environment instrumentally, that is, simply as a means to human ends and values. Thus they take account ultimately only of human interests and concerns.

Nonetheless, there are a number of ethical arguments, justifying better dealings with the environment, open to shallow positions. Most obvious are those that appeal to human welfare and to the conservation, preservation or protection of the natural environment or things in it of interest to humans. The two most popular forms of these arguments are prudential arguments and instrumental arguments. Prudential arguments hold that human well-being depends on the well-being of the environment, or even that human survival depends on the survival of the environment, and it is, therefore, in the best interest of humans to preserve *their* environment. Instrumental arguments construe nature and the environment simply as means to human ends and values. Instrumental value or consideration is attributed to an item that is in some way of value to or useful to another.

PRUDENTIAL AND INSTRUMENTAL ARGUMENTS

The adequacy of shallow environmentalism turns upon these arguments. We turn accordingly to assessing them. Prudential arguments are pleas for prudent behaviour and are typically grounded in human interests. These interests may be quite oblique, amounting for instance to merely thinking that something (valuable) is there. Prudential arguments are arguments encouraging humans to exercise wisdom, but mainly the wisdom of protecting human interests. These arguments have a powerful appeal because they pander to human interests, but they do not acknowledge the value of the environment or parts of it, except so far as they promote human well-being or survival.

There are a variety of instrumental arguments. The Australian philosopher and historian of ideas, John Passmore lists economic, scientific, recreational, renewal, aesthetic, and posterity varieties. To take the last argument first, the 'Posterity Argument' holds that the present generation has a duty to hand over the world "to our successors in a better condition".[1] Part of handing over the world in a better condition (from an ecologic, as opposed to the economic, perspective) is not to destroy species or let them perish through contributory negligence and thereby deprive posterity of the opportunity to observe, study, and otherwise have the experience of the diversity enjoyed by the present generation. For example, this is the last generation with a chance to preserve pristine rainforests. If this generation does not do it, no other can. With a 'Posterity Argument' Passmore is not calling for passive restraint on the part of humans in their dealings with the environment. For him, handing over the world in a better condition means "transforming the world into a civilised state".[2] Betterment through transformation *may* mean the preservation of some biospheres, but with the major responsibility to other humans, it could mean the destruction of parts of the environment considered harmful or 'uncivilized'. Civilization has too often been on a collision course with environmentalism.

Yet, Passmore attempts to soften the harsh impact of this line of reasoning by a second argument: the 'Gene Pool Argument', or as another Australian philosopher, William Grey labels it, the 'Silo Argument'.[3] As humans utilize the plant and animal resources of the world they modify them through cultivation and domestication, but a modification for one purpose, say breeding cattle to produce more beef or greater drought resistance, can cause undesirable side effects, say less hybrid-vigour. Therefore, it is in humanity's own interest to keep wild stocks as a reservoir or gene pool.

Passmore holds "a purely economic argument will suffice to establish at least a prima facie case against the clearing of wilderness [and] the destruction of species".[4] But obviously the 'Gene Pool' or 'Silo Argument' is reasonable only for preserving sample wild counterparts of domesticated species that produce meat, milk, fur, or other economically valuable products. Also it could be objected that a prima facie case does not hold if it is more economic to destroy a wilderness or species for immediate gain. Poaching elephants for their ivory and rhinoceroses for their horns have placed both on the edge of extinction.

To protect species that are not currently held to be valuable, Passmore appeals to the 'Rare Herb Argument'. "A species often turns out to be unexpectedly useful, a tropical plant to contain pharmacologically valuable substances".[5] So Grey claims of this argument that although prudence counsels that we preserve areas that can contain rare herbs and that although environmentalists often use the argument as a lever, "The rare herb argument is a sound argument in favour of preservation on the grounds of *prudence*, but it is not, on the face of it, a *moral* argument".[6] Even if it is effective, it certainly need not be a moral argument. But prudence has not prevented the destruction or near destruction of rainforests and the myriad of species that inhabited them. In Australia alone 18 species of birds and mammals and 78 species of plant have become extinct by direct or indirect human intervention since 1788.[7] Hopefully the 'Rare Herb Argument' could be applied to argue against the obliteration of additional species. But it is hardly adequate. For it says nothing of the numbers, conditions, or manner under which these reservoirs are kept, and it could permit the decimation of these species to zoo populations, breeding colonies, or token specimens.

A complementary argument to the 'Rare Herb Argument' is the 'Laboratory Argument'. The 'Laboratory Argument' states the obvious thought that unless possible sources of pharmacologically valuable substances are preserved for experimentation then the diversity of humanity's environmental laboratory is diminished. As with the 'Silo Argument', the thrust of the 'Laboratory Argument' is that biological modification and experimentation by humans requires natural species as norms and fresh resources. Grey states it in a way of which Passmore would surely approve:

> This is the argument that wilderness areas provide vital subject matter for scientific inquiry which provides us with an understanding of the intricate interdependencies of biological systems, their modes of change and development, their energy cycles, and the sources of their stabilities. If we are to understand our own biological dependencies, we require natural systems as a norm, to inform us of the biological laws which we transgress at our peril.[8]

Duly followed through, this argument will support significantly larger natural reserves and regions of biodiversity than are now contemplated, or left, in many countries.[9] But, on shallow approaches, too much depends (wrongly) upon sufficient interest and good-will from humans, not presently conspicuous. In addition to the environment providing material resources, Passmore also holds, "that there is refreshment as well as enjoyment to be found in wandering through wild country. (Not only recreation but re-creation; it renews one's sense of proportion)".[10] When duly elaborated, this can be split into two arguments: the 'Recreation Argument', where humans use wilderness as their gymnasium and the 'Cathedral Argument' where humans draw spiritual revitalization from wilderness.

An instrumental argument upon which Passmore builds much of his overall case for a stronger adherence to established principles and ethics, rather than the development of a new ethics, is the 'Aesthetic Argument'. Passmore holds that, "to justify action against the beauty-destroying polluter; at most what is needed is a strengthening of existing moral principles".[11] The 'Aesthetic Argument' is not limited to protecting the environment against polluters. As another Australian philosopher, H.J. McCloskey, who has written extensively on environmental philosophy, observes, "we hold the beautiful and mysterious in nature in awe and wonder, see much as precious, and view with moral shock and even moral outrage the wilful destruction of beautiful species".[12] The 'Aesthetic Argument' conveys the concept of 'beauty-destroying', which Passmore finds is a "more explicitly Western tradition that it is wrong unnecessarily to destroy – a principle embodied in the concept of 'vandalism'".[13] This argument places the moral onus on the destroyer. It is intended to protect the environment from "'wanton' destruction where no defence of the destruction can be offered".[14]

For the most part, these sorts of arguments justify preserving at best tiny bits and pieces of the natural world: a few relics, reserves, monuments, and so on. While diversity is certainly an important qualitative consideration, *quantity* matters as well. What is required, for example, is not just wilderness but substantial wilderness (as *other* arguments help show). Worse, many of the natural reservations may be but temporary; they are liable to fall for 'higher' human values as human numbers and demands grow.

Chapter 2

ON PASSMORE'S SHALLOW ENVIRONMENTAL ETHIC

Beyond these specific forms of shallow argumentation, Passmore presents over-arching, embracing attitudes about the connections between ethics and treatment of the environment. Passmore's position is a hallmark among shallow environmental ethics. Passmore's work in this area, particularly in *Man's Responsibility For Nature,* maintains that what is needed to preserve the environment is a stricter adherence to current forms of ethical consideration and not the development of new systems. He concentrates on four environmental problems: population growth, pollution, resources, and preservation of species. Three of these four issues reflect components in the standard environmental impact formula. Passmore attacks the variables with a three pronged attack, according to British philosopher, Mary Midgley, although she doubts that his prongs always point the same direction. According to her, the prongs are:

1. *Logical.* The demand for total change is incoherent, simply does not make sense – as he says, "you cannot change your morality as you can change your coat".

2. *Pragmatic.* It is no *use* appealing to standards, which people do not in fact already recognize.

3. *Moral.* If you go against existing standards, what you do is wrong, so your last state is, one way or another, worse than your first.[15]

Some aspects of these prongs will become evident, although, as all are questionable, they will not be relied upon heavily for an explication of his position.

Passmore begins by laying out some of the history of Western thinking as it is particularly relevant to treatment of the environment. The foundation of Western attitudes towards the environment is derived from the Stoic-Christian traditions adopted.[16] Nature is alien:

> the Stoic-Christian tradition has insisted on the absolute uniqueness of man, a uniqueness particularly manifest, according to Christianity, in the fact that he alone, in Karl Barth's words, has been 'addressed by God'

.... If nature, on that view, is not wholly strange, this is only because it has been created by God for men to use. Animals and plants can for that reason be assimilated, at least in certain respects, to the class of tools, dumb beasts but none the less obedient to men's will.[17]

The view that everything existed for man's use brought about an attitude towards nature that it was not something to respect, but something to utilize and associated with this concept of nature "was a particular ethical thesis: that no moral considerations bear upon man's relationship to natural objects, except where they happen to be someone else's property or except where to treat them cruelly or destructively might encourage corresponding attitudes towards other human beings".[18]

From his precis of the environmental attitudes embedded in Western traditions, Passmore develops two varieties of a Dominion thesis that have been conveniently summarized by Australian environmental philosopher, Val Plumwood:

> Passmore distinguishes what are essentially two varieties of the Dominion Thesis. Both take the view that humans are entitled to manipulate the earth, its ecosystems, and all its non-human inhabitants for human benefit, which is the essential and basic element of the Dominion Thesis. But the stronger thesis which Passmore calls the Despotic view, holds that the earth and its contents were created for, exist entirely for, the benefit of man, who can control its systems with very little effort or need for skill or understanding. 'Nature is wax in man's hands' and there are virtually no constraints on man's relations with nature. This view has been encouraged by some religious doctrines in the past.
>
> The second, which we might call the Responsible Dominion View, takes the line that while it is, of course, permissible for man to manipulate the entire earth exclusively for human benefit, it is usually difficult for humans to do so successfully and without unwanted ill-consequences for themselves. Natural processes exist independently of humans and do not serve them. Successful manipulation requires skill, knowledge and understanding of natural processes, and taking careful account of the likely consequences of actions on other humans. To this extent, nature must be 'respected' while being made to serve human interests. These factors impose substantial constraints on man's permissible actions with respect to nature.[19]

These two variations reduce to a single ethical view that Plumwood calls the Dominion Assumption. As described, the Dominion Assumption, and in particular the Responsible Dominion View, has lead supposedly to an attitude of attempting to understand the laws of nature and of transforming nature through technology. How much 'understanding' there is or how much wisdom has been used in applying technology based on that understanding is dubious:

> nature is not a passive recipient of human action. When they operate upon it, they are affecting existing modes of interaction as distinct from merely modifying a particular characteristic. Nature, in other words, does not simply 'give way' to their efforts; adjustments occur in its modes of operation, and as a result their actions have consequences which may be as harmful as they are unexpected. That is the force of the dictum, now so popular amongst ecologists, 'it is impossible to do one thing only'.[20]

More importantly from an ethical perspective, there is an internal tension in the Dominion Thesis – ultimately to dominate nature means to destroy it and to destroy it means to destroy humans. Passmore admits, "Western metaphysics and Western ethics have certainly done nothing to discourage, have done a great deal to encourage, the ruthless exploitation of nature".[21] Yet, despite this ruthless exploitation of nature, there resides in Western traditions, Passmore claims, "the tradition that sees [man] as a 'steward', a farm-manager, actively responsible as God's deputy for the care of the world.... The tradition of 'stewardship' – never strong but persistent – dates back to the post-Platonic philosophers of the Roman Empire".[22]

The tradition of stewardship is derived from a hierarchical arrangement God:Humans:Nature. Under this arrangement God put humans on the earth in order that they should look after it, i.e., nature. While humans served as stewards, the ultimate ownership of the earth was never for a moment in doubt. Passmore argues that it is still the role of humans to maintain nature, not to wantonly destroy it. Nevertheless, he wants to maintain this 'stewardship' role without the first element in the hierarchy – God. Plumwood has summarized Passmore's position on 'stewardship':

> according to at least some versions of the Stewardship position, humans do not have absolute title to the earth but are merely Stewards for God,

and have obligations to care for the plants and animals of the earth because God cares for them, even if humans do not. Thus they are not entitled to manipulate the earth exclusively for their own benefit. Even if the directly theistic interpretation of Stewardship is avoided, and God is taken as a personification of value, what follows is that the value of natural items may not simply reduce to their value for human interests, and this contradicts the Dominion Assumption. It is only if God is taken as a super-human (so that concern for these non-human items can be reduced to a matter of human or super-human interest) that the Stewardship view will sanction the Dominion Assumption.[23]

The thrust of his advocacy of stewardship and his ethical project is to adhere to and to reinforce those elements of Western traditions that allow consideration for the environment and thereby to construct a more respectful and responsible attitude and ethic towards the environment without losing sight of those elements in Western tradition that have made it great, if not unique.

> A morality ... is not ... the sort of thing one can simply conjure up. It can only grow out of existing attitudes of mind, as an extension or development of them, just because, unlike a speculative hypothesis, it is pointless unless it actually governs man's conduct. But it may be true that in fact men's attitudes are already changing, that the 'new morality' would be a natural outcome of a change that is already in process, which can now be hastened by exhortation or argument [24]

As already emphasized in the discussion of the 'Aesthetic Argument', Passmore holds that one principle (or as he calls such principles – one 'seed')[25] residing in Western traditions that can be used to actually govern human conduct is the "tradition that it is wrong unnecessarily to destroy – a principle embodied in the concept of 'vandalism'".[26] While this is no doubt a fine principle, Passmore goes on to admit that it has not been, to his knowledge, much emphasized by Western moralists. Yet it places the moral onus on anyone who wantonly destroys, where no defence of the destruction can be offered. "This is particularly so when, as in the case of species, the destruction is irreversible".[27] The trouble, for Passmore, is that these seeds are liable to produce fruit exceeding the Dominion framework, as when, with a more thoroughgoing or deeper criticism of vandalism, they involve values beyond human interests and

71

concerns (for ask *how* Passmore justifies his use of an anti-vandalism principle where these concerns are unaffected: briefly, vandalism presupposes *value* of what is destroyed). Furthermore, the principle does not work against, what is as insidious, creeping economic destruction.

In appealing to 'seeds' such as the wrongness of wanton destruction, Passmore tries to avoid attacks on those portions of Western culture that he considers valuable, in particular on Western science, "perhaps the greatest of man's achievements".[28] Western science is under attack because it is seen as a major part of the problem of treating nature disrespectfully. Passmore rejects attacks on science as a reversion to mysticism – a cheap rejection, since the enterprise of science is under attack for several other reasons. He argues that Western traditions deliberately rejected regarding nature "as having a 'mysterious life' which it is improper, sacrilegious, to try to understand or control".[29] Instead, Western traditions turned to science, which "in contrast, converts mysteries into problems, to which it can hope to find solutions".[30] This does not mean that science cannot be reformed to be more in keeping with promoting the civilization of nature, without wantonly destroying it. Indeed, Passmore holds that science should go hand-in-hand with civilizing nature as humans, "uniquely, are capable of transforming the world into a civilised state; that is their major responsibility to their fellow-man".[31] The idea of 'civilizing' the world, that is transforming it from its material state to a human artefact runs in diametrical opposition to the idea also proposed of leaving part of the world in a natural, uncivilized state, and also contradicts the very notion of an environmental ethic. An environmental ethic does not have to insist on nature remaining totally 'untouched' or 'unspoilt', but it does have to insist on nature not being totally transformed to a civilized state. If Passmore's environmental philosophy is truly one intended to reverse the vandalism and prevent the wanton destruction he deplores, then there is an internal tension between it and his push for "transforming the world into a civilised state".

"The main philosophical issue raised by conservation, as Passmore sees it, is the question of our moral obligations to posterity".[32] Humans should civilize nature and pass it on in a more civilized, yet not abused form, to future humans. Humans are all that basically count. "In these attitudes, Passmore is firmly within the European tradition, with its emphasis overwhelmingly on the value of human achievement and artefacts rather than on the value of natural areas".[33]

The question remaining is: What is humanity's responsibility to nature? It should be noted that the title of Passmore's book is *Man's Responsibility For Nature*. He is not claiming that humans have a responsibility **to** nature. "The basic unsoundness in Passmore's case lies in his tendency not to count as real any problem which is not a problem for human interests; if so, it is unsurprising that the man-centred ethic which he favours is found to be adequate to their solution".[34] In subsequent writings, Passmore concedes that environmental problems such as pollution "can affect nonhuman species in a way that makes it difficult to invoke existing moral principles to condemn the pollutor".[35] Thus are some of the fundamental weaknesses of shallow environmental positions exposed. Firstly, environmental problems that do not connect with human interests are not candidates for ethical consideration. So, for instance, it remains ethically proper to continue to dump wastes off-shore in Sydney until enough people become ill from eating the fish or swimming in the water. Secondly, human interests change; some are quite fickle, or merely trendy fashions. What is a candidate for ethical consideration today, may not be considered so tomorrow, simply because it is no longer an object of human consideration. But ethics are not like that; principles cannot fluctuate so freely.

A major strength of shallow environmental argumentation is supposed to be that it appeals to humans to save the environment because their survival, happiness, diversity of experience, and well-being among other things depend on it. An appeal to 'enlightened self-interest' is usually taken to be a very powerful argument for environmental con-sciousness and conservation, sometimes, by those firmly locked into the Enlightenment ideology, as the most powerful argument for conserva-tion, sometimes, even more erroneously, as the only argument. At least, it is assumed that once humans have been shown that it is in their interests to save the environment, they will do so, because humans are highly adaptive and because it is inconceivable that – faced with a choice between extinction and survival, for instance – they should choose extinction. However, most humans miss seeing that saving the environment is a choice between survival or extinction of their species. Troubles do not end there, with humans failing to recognize, or to politically effect, their enlightened interests, or even their long-term group interests.

On the one hand, the anthropocentric argument – that appealing to human survival and so on – is weak. By the time a wider long-term perspective has overcome short-term interests it will be too late. For instance, conservationists find it amazing that wood chippers fail to see that the jobs they are fighting so hard to keep now will be gone with the forest in one, two or five years' time when they have clear felled the forest. For the most part, industrialized societies patch and repair some of the worst environmental damage rather than prevent it or promote a healthy environment. Fish disappear from the Thames, the Hudson, or the Seine, and much later a concerted effort reverses the rot to the extent that some tough fish reinhabit it and pollution levels are lowered somewhat.

Environmental preservation or conservation is a long-term group issue, but most humans live on a local individualistic perspective. Even survival of humans as a species is of little consequence to most humans as individuals. The short-term arguments on which their lives depend will precede and obscure long-term arguments on which the survival of the species depends. Most individuals are concerned with their lives and the lives of their children. Selves, spouses, sons and daughters are vital concerns – not posterity more than three or four generations down the line (where their selfish-genes are already beginning to disappear in a group genetic mix). Similar considerations help reveal why Passmore's own sketchy chain-of-interest ethics is so unsatisfactory; it does not allow properly for wider concerns and assessments where chains peter out, or fail to reach, or conflict. As well, of course, any such ethics violates ethical impartiality requirements, by favouring immediate family.

On the other hand, the real power of the anthropocentric argument may lie in its similarity to a television family affairs soap opera. A television soap opera grabs the viewer's interest at a personal and emotional level, and may hold interest even if it does not inspire deeper thought into problems or issues. If the viewer's attention is grabbed, then the viewer at least sees the problem. The viewer is at least aware that there is a problem. Environmental awareness is the first step.

Shallow argumentation – especially instrumental and prudential arguments – accords some, though insufficient value, merely instrumental value, to the environment and items in it, without including them in the moral community or giving moral consideration to them for their own sake. Another form of argumentation, such as that of extension arguments, argues for the recognition of the value of the environment and items in it for their own sake and includes them in the moral community, as will now be explained.

Intermediate Environmental Ethics

All the arguments so far considered maintain that the environment, or at least certain significant parts of it, are worth saving, but worth saving because humans need them or have an interest in them or similar. Instrumental and prudential arguments recognize that humans need the environment more than the environment needs humans. But the conservation, preservation or protection of the environment depends on humans, their interests and so on. What distinguishes intermediate arguments is that they deny that humans are the only thing of value, that is, they deny the 'Sole Value Assumption'. One important variety of intermediate argument in environmental ethics consists in extension arguments. Two examples of extension arguments are Aldo Leopold's Land Ethic and the 'Argument from Marginal Cases'. What these arguments have in common, in addition to rejecting the 'Sole Value Assumption', is that they take established ethical frameworks and extend them beyond the human realm.

A LAND ETHIC

One early example of an intermediate environmental ethic has had considerable influence in Australia and elsewhere, and is the prototype and perhaps best known example of an environmental ethic – Aldo Leopold's Land Ethic. Leopold was an American forester, later an ecologist, who has become an ecosaint of the environmental movement worldwide. The beauty of his Land Ethic is its simplicity. The extension is based on two fundamental principles. First, there is his primary ethical principle, which is presented in his work as the maxim, "A thing is right when it tends to preserve the integrity, stability, and beauty of the biotic community. It is wrong when it tends otherwise."[1] Thus he recognized that items in the natural environment, such as a biotic community, have value-in-themselves as well as or despite any value they may have for humans. The primary principle (which enjoys initial plausibility as a first approximation in characterizing *right*) includes denial of the 'Sole Value

Assumption'. His second principle is his call for an extension of ethical consideration to the land, or more broadly and accurately, extending the ethical community to include the ecosphere. Leopold sought to make the ethical and biotic communities co-extensive. His extension principle is summed up as, "The land ethic simply enlarges the boundaries of the community to include soils, waters, plants, and animals, or collectively, the land".[2]

Hitherto, so extensionism maintains, humans had been ethically or morally special in that they are both the objects of moral concern and the frame of reference for determining moral relevance. The notion of 'simply enlarges' implies that currently accepted principles and norms for appropriate and inappropriate behaviour would remain unchanged or relatively unchanged while the community would be expanded to include the non-human world. Humans would no longer be the sole objects of moral concern and the ethical community would be radically redefined to make it co-extensive with the ecological community. This is precisely the move in Leopold against which Passmore argues. Passmore argues that it may be true that humans are part of the biotic community, but that does not make the community an ethical community:

> Ecologically, no doubt, men form a community with plants, animals, soil, in the sense that a particular life-cycle will involve all four of them. But if it is essential to a community that the members of it have common interests and recognize mutual obligations then men, plants, animals and soil do not form a community. ... In the only sense in which belonging to a community generates ethical obligation, they do not belong to the same community.[3]

At first glance it would appear that Passmore and Leopold agree on the basic ecological fact, but differ on how to value this fact. But Leopold would not accept Passmore's limited meaning of an 'ecological community'. The ecological community includes the concepts of interdependency of the members and elements of the community, and the greater dependency of humans (and other lately-evolved omnivores and carnivores) on the community. Leopold accepts the concept of the ecological community as both environmentally descriptive and ethically prescriptive. Unlike Passmore who holds that humans should use science to understand the laws of nature and technology to transform nature,

Leopold holds that understanding nature includes valuing nature. Moreover that understanding should direct, if not dictate, how it is valued. Furthermore, Leopold would reject what seems to be a conflation of a sociological sense of community (i.e., composed of one species) with an ecological sense (i.e., composed of a diversity of species). That is, the ecological community is not simply a loose collection of members but an association of mutual dependency. Leopold does not say that the ethical obligations have to be mutual in the ethical/biotic community. Rather he indicates, unfortunately only vaguely, what J. Baird Callicott (following Leopold's lead) later makes explicit regarding such a land ethic, that it is "a distinct ethical theory which provides direct moral standing for the land (in Leopold's inclusive sense)".[4] An environmental ethic starts with the ecological or environmental community and that community forms the ethical community. The ethical relationships within the community are based on the relevant characteristics applicable to members of the whole community rather than those possessed by highly competent members of one species, namely, humans. However, it is only the relationships between humans and nonhumans that are changed.

Passmore imports a restricted sense of obligation to prevent the ethical community, as he narrowly sees it, from becoming co-extensive with the biotic community. Passmore's objection suggests that beyond enlarging or redefining the ethical community to which direct moral concern or consideration is due, it also makes humans the moral watchdogs for interactions among nonhuman members of the community. But, because other members of the community are incapable of upholding their moral obligations within the community, it is not a community. The stipulation presumed is known as the Reciprocity Assumption. "This is the assumption that if two individuals stand in a moral relationship, then both individuals must be fully-fledged moral agents. A moral agent, I assume, must be capable of exercising reflective rational choice on the basis of principles".[5] Simply put (in deontic terms) reciprocity means that if individual **A** has a duty or obligation to individual **B** concerning some action or relationship, then individual **B** should have a correlative duty or obligation to individual **A** with regard to that action or relationship. However, it is easy to find exceptions where reciprocity does not apply and further to extend these exceptions to relationships like a land ethic. One such exception is the relationship between adult humans and infant humans. Adult humans may have duties or obligation to their or others' infants, but those infants are not expected to have the same, or for that matter any, duties or obligations to the adults. The same applies to many other classes of humans, such as retarded persons. The point to be emphasized from this is that in a moral community not every individual must be a moral agent: in the words of William Grey, "although I do not think that we could legitimately speak of a moral relationship obtaining between two individuals *neither* of which is a moral agent, is there any good reason for supposing that *both* individuals must be moral agents?".[6] While this is a rhetorical question, no doubt should be left as to the answer. American moral philosopher, James Rachels spells it out. One individual may "lack characteristics necessary for having obligations; but they may still be proper beneficiaries of our obligations. The fact that they cannot reciprocate, then, does not affect our basic obligations to them".[7]

Furthermore, the 'common interests' within the ecological community are different from the common interests of one species (any one species), which Passmore apparently fails to realize. Members of one species have interests in common that they do not have in common with other species, but they also have interests in common with other species. Ecological communities are composed of mutually dependent, but heterogeneous entities, and are spanned by families of interests.

The human position within the community is special only in that (except for the contested exception of God) humans supply the only presently known moral agents as well as constituting the species posing the single greatest threat to the rest of the community, and thereby to itself. It is human behaviour within and treatment of the community that can be and *has* to be moderated and monitored. As Callicott puts this position, again following Leopold's lead:

> We wish to 'enlarge the boundaries of the community' to include nonhuman natural entities as beneficiaries of moral obligation, but without imposing upon them mutual or reciprocal obligations, duties, or moral limitations, which it would be impossible for them to bear and absurd for us to suppose that they might.[8]

In addition to its simplicity and austere beauty, Leopold's Land Ethic has a further appeal. It is a logical next step following on the sorts of shallow environmental argumentation exemplified by Passmore and the instrumental and prudential arguments. Yet as appealing as it may seem to simply enlarge the boundaries of the moral community, it is not as simple as Leopold makes it appear. To "simply enlarge" the ethical boundaries "by the process of 'extension' perpetuates the basic presuppositions of the conventional modern paradigm, however much it fiddles with the boundaries".[9] In other words, while there may be some advantages for the environment through extending the moral community, this extension carries with it all the disadvantages as well. For instance, it is still primarily a human-centred ethical system. It seems that human convenience could still override consideration for the environment for comparatively trivial reasons. Schweitzer's criticism still applies – an extension of previous ethical theory is an extension of an ethical relationship of one human to another (whereupon arise all the problems of avoiding stepping on beetles, and indeed of managing day-to-day living at all!) To put it paradoxically, it is not until humans accept their dependency on nature and put themselves in place as part of it, not until then do humans put humans first (and if they have really seen what is involved, they won't always). This is the paradox of such an environmental ethic.

Leopold's ideas are, however, subversive and constitute a landmark in the development of a new position. Analytically, tearing his basic premises out of context has done him a disservice. The new position he must have desired was not an extension of the problems of an anthropocentric ethic, but an ecological consciousness expressed in an ethic among other ways:

> This extension of ethics, so far studied only by philosophers, is actually a process of ecological evolution. Its sequences may be described in ecological as well as in philosophical terms. An ethic, ecologically, is a limitation on freedom of action in the struggle for existence. An ethic, philosophically, is a differentiation of social from anti-social conduct. These are two definitions of one thing.
>
> There is as yet no ethic dealing with man's relation to land and to the animals and plants which grow upon it. Land ... is still property. The land relation is still strictly economic, entailing privileges but not obligations.

The extension of ethics to ... the human environment is, if I read the evidence correctly, an evolutionary possibility and an ecological necessity. ... Individual thinkers since the days of Ezekiel and Isaiah have asserted that the despoliation of land is not only inexpedient but wrong. Society, however, has not yet affirmed their belief. I regard the present conservation movement as the embryo of such an affirmation....

Ethics are possibly a kind of community instinct-in-the-making. All ethics so far evolved rest upon a single premise: that the individual is a member of a community of interdependent parts.... The land ethic simply enlarges the boundaries of the community to include soils, waters, plants, and animals, or collectively: the land.

In short, a land ethic changes the role of *Homo sapiens* from conqueror of the land community to plain member and citizen of it. It implies respect for his fellow members, and also respect for the community as such.[10]

Leopold was not a philosopher and some of his ethical remarks, principles, and solutions are perhaps philosophically naive; but his Land Ethic is an important transition from shallow environmental ethics to a new appreciation of the necessity to acknowledge ethical consideration directly *to*, rather than indirectly *for* the environment. It illuminates the paradox of environmental ethics and highlights directions in which some answers to the environmental crisis lie. As an example of this, he was aware of the need to integrate environmental and economic concerns to produce a system that was sustainable for other species as well as humans, "I believe that many of the economic forces inside the modern body-politic are pathogenic in respect to harmony with land".[11]

What is required is to shift the frame of reference for ethical consideration from humans to the biotic community. With a proposal like Leopold's the biotic community or the environment or nature not only would be the object of moral concern but also the characteristics, attributes, qualities or excellences of the biotic community or items in the environment would be used as determinants of moral relevance. Human survival, like the survival of other species, remains a central concern without humans remaining the central focus. Humans' survival is no less an ethical concern than it ever was, but the preservation of humans is through and not at the expense of the preservation of the ecological community. The traditional assumption of the dominant Western

paradigm of human supremacy over the rest of creation – an environmental 'might makes right' assumption – is rejected. And human
insensibility and insouciance towards the rest of the biotic community
is supplanted with due and direct regard.

The second application of an extension argument is the 'Argument from Marginal Cases', sometimes called the 'Argument From
Human Analogy and Anomaly'.[12] It is one of the mainstays in Peter
Singer's case for animal liberation. Unlike the philosophically naive
beginnings of Leopold's Land Ethic, philosophers have played an
important part in the development of animal liberation. In particular,
Singer, now Professor at Monash University's Centre for Human
Bioethics, inspired the movement worldwide with his stand, both
practical and theoretically underpinned, against the callous treatment of
nonhuman animal species. Singer's theory, an application of Bentham's
already extended utilitarianism, arose within the setting of the British
animal welfare and animal rights movement. Although it has aroused
considerable hostility, Singer's stand has drawn considerable public
attention, not the least of which has been the rise of activist organizations
such as Animal Liberation in response to it.

ANIMAL LIBERATION.

Before going on to elaborate on philosophical theory expressed and
implied in Singer's *Animal Liberation* and to define speciesism, it should
be re-emphasized that the Animal Liberation Movement in Australia was
founded on a philosophical text, more explicitly, on a text that is part of
the corpus of environmental ethics literature.

In Australia, Animal Liberation began with the following four
objectives:

1. To abolish man's speciesist attitude towards animals;

2. To conserve wildlife by ensuring its habitat remains undisturbed;

3. To promote a conservation policy which entails mercy and
   protection of animals, instead of exploitation purely for man's
   benefit;

4.   To carry this out according to the philosophies expressed and implied in Peter Singer's book *Animal Liberation*.[13]

As will be explained later, the Animal Liberation Movement is a case in which theory and practice have reinforced each other. Like Leopold's Land Ethic, one of the appeals of Singer's argument and theory is its simplicity. The 'Argument from Marginal Cases' is a call for consistency and equality, or better expressed, for uniform and similar treatment. The 'Argument from Marginal Cases' is an ethical argument for such consistency of the form: if entity **A** is given certain moral considerations or treated in a certain manner because it has some given characteristic(s), then entity **B** having the same or similar characteristic(s) should be given the same moral consideration or treated in the same manner regarding that characteristic or characteristics. However, it contains the distinctive elements of arguing from non-paradigmatic cases and of placing the moral onus on those opposed to ethical consideration for nonhuman animal species to find a morally relevant distinction to deny the call for consistency. Jan Narveson tries to explain why the argument is called the 'Argument from Marginal Cases': "Humans differ from animals in having more sophisticated intellectual and emotional equipment, but they are the same in having the capacity to suffer and enjoy. And we believe that this latter capacity is the source of rights independently of the other capacities; for we find that we do not believe that imbeciles or infants may be used just as we please, and yet they are as little, or even less, possessed of the more sophisticated capacities than many animals".[14] This is the sort of argument that was used in helping to abolish slavery and, more recently, to secure civil rights for blacks and equal opportunity for women as well as animal liberation. Unlike Leopold, Singer develops more complex arguments to back up the 'Argument from Marginal Cases'. These are drawn, though inessentially, from a utilitarian setting.

Although throughout Singer is a no-nonsense utilitarian, a follower of Jeremy Bentham, utilitarianism can be cleanly shorn from most of his arguments for animal liberation. He grounds *Animal Liberation* upon Bentham's premise, "the question is not, can they *reason*? nor, can they *talk*? but can they *suffer*?" – a rhetorical question to which non-utilitarians can give the same evident answer.[15] While vestiges of René

Descartes' position that animals are biological machines and do not feel pain, still linger, few would doubt now that nonhuman animals do feel pain and experience pleasure. Indeed, much medical and psychological experimentation (the excesses of which Singer argues against) is based on the assumption that animals are in given respects analogous or homologous to humans in their capacities. If the capacity to suffer of some or many nonhuman animal species is analogous or homologous to humans (and there is little doubt that it is, even though isolated sceptics imagine conclusive proof is lacking), then it is easy to see the powerful appeal of the 'Argument from Marginal Cases'. That argument undermines speciesism.

Speciesism, a term coined by Richard Ryder, "is a prejudice or attitude of bias towards the interest of members of one's own species and against those of members of other species".[16] Those who have overcome this prejudice are often compared to those who try to prevent needless and unjustifiable human suffering, such as those working against racism and against apartheid or for women and for the environment. "The chief difference between them and those whose who work exclusively for human welfare is that the animal liberationists have pushed the boundaries of their concern back one stage further. They see nonhuman animals as another oppressed group, suffering from blatant exploitation by a species that has unlimited power over other species and uses this power for its own selfish interests".[17]

Singer's book is directed against the callous treatment meted out by technological societies to nonhuman animal species. Nonhuman animal species are used in repetitious and, in many cases, needless experiments to test the toxicity or irritability or some other factor of cosmetics, cleansers and detergents, and other household products. Commonly the results are already known from previous experimentation on similar or identical substances. Rabbits are submitted to the Draize test. A substance to be tested for properties such as levels of irritants is placed in the eyes of rabbits, who do not have tear glands to wash away the substance.

Rats and other species are submitted to the LD 50 test. LD stands for lethal dose. The animals are force fed or injected with the substance to be tested for toxicity until 50 per cent of the test group dies. If toxicity is low it is often the volume of substance rather than its poisonous properties that kills.

Chickens are kept as 'production units' in battery cages to produce eggs. Four or five hens are confined to a 16 by 18 inch cage and as many as nine hens in a cage 18 by 24 inches. Mortality among layers housed in these conditions can run as high as 23 per cent. Under these conditions they develop what are know as 'vices' in the factory hen business. These 'vices' include pecking each other to death, so the hens are debeaked. Intensively farmed pigs are given as little as 7 square feet per pig. Sows are immobilized during pregnancy and lactation by attaching them to the floor of their stalls with holding frames. Often piglets are weaned within a week to ten days so that the sow can be reimpregnated, thus making it possible for each sow to have on the average 2.6 litters per year. The sows become bored, but eat less so they are more economical. It is not only what is happening, but to how many animals. It is not only the needless suffering, but its magnitude. Each year in the United States alone 70 million animals are used in experimentation and "5 billion animals spend their entire lives in factory farms".[18]

In addition to being fluently argued, Singer's theory has two distinct advantages as an extension ethic over Leopold's Land Ethic. A first advantage is that animals, particularly charismatic mega-vertebrates, are easier to identify with and thus more readily considered for inclusion into the moral milieu. After all, almost everyone who has experienced pain wishes to avoid feeling it gratuitously, and in most cases inflicting it gratuitously. Animals are also easier to identify with because they are more like humans. It is easier to empathize with a deer in a field, than the field the deer is in. A second advantage develops from this last point; the objects of the arguments are more readily available to most people and it is often easier to take action concerning them. It is easier to buy free-range eggs, than to prevent the destruction of a rainforest or even a small part of it (however purchase of rainforest timber can be eschewed, etc.). Of course, it also has certain major disadvantages, like the very shabby treatment of major environmental issues, such as wilderness, forest destruction and human overpopulation.[19]

Several of Singer's arguments are based on the principle of equal consideration of like interests. To understand these arguments, it is necessary first to understand several of his terms and premises. One is interests, now a vexation. Two morally relevant senses of interests can be distinguished as Stanley Benn did:

> the sort of interests that organise or give a consistent direction to otherwise diverse activity - like an interest in music, or football, or making money. ... But we can say that something is in someone's interests, where what is meant is only that it is conducive to his well-being.[20]

In other words, a distinction between the sense of 'in someone's interests' and 'someone's interest in'. It is conceded that nonhuman animals have interests in the sense of 'in someone's interests', but it is the second sense that often draws attention in debates over moral consideration, especially rights, for nonhuman animals. It is the second sense that is thought by many to carry real moral significance. McCloskey, for one, takes it as a foregone conclusion that nonhuman animals do not have interests in the sense of 'someone's interest in', because "the concept of interests has this evaluative-prescriptive overtone".[21] Singer argues that interests in both senses are important and that interests of nonhuman animals warrant consideration for their own sake and that humans have direct duties to members of other species.[22] Any living thing that evidences interests is worthy of moral consideration.

Another of his premises concerns sentience, which has already been discussed. Singer uses sentience as a shorthand for the capacity to feel pain and experience pleasure or the capacity to suffer and enjoy. This is a basis for the 'Argument From Marginal Cases' used. Combining the elements of sentience and interests produces the following sort of argument – a cat and a human infant have equal interests in avoiding unnecessary pain, even though the cat may feel the pain of a slap on the face, say, more than the infant. Singer recognizes that there are morally relevant differences between humans and nonhumans:

> there are important differences between humans and other animals, and these differences must give rise to some differences in the rights that each have. Recognizing this obvious fact, however, is no barrier to the case for extending the basic principles of equality to non-human animals.[23]

Singer holds equality to be an ideal, rather than an assertion of fact:

> Equality is a moral idea, not an assertion of fact. There is no logically compelling reason for assuming that a factual difference in ability between two people justifies any difference in the amount of consideration we give to their needs and interests. *The principle of the equality of human beings is not a description of an alleged actual equality among humans: it is a prescription of how we should treat humans.*[24]

Even as a prescription, the principle is problematic (consider, e.g. triage situations). In any event, the principle of equality, which has something right about it, is extended to apply to relations between humans and nonhumans in the same manner as it applies between humans. Singer proposes the extension of ethical consideration to the interests of other species to give them a equal opportunities with humans. But while Singer's proposal does extend moral consideration from humans to all sentient beings – though only them, thus trading human chauvinism for *sentient* chauvinism – his move does nothing to change a damaging utilitarian orientation. Nonetheless, the importance of extending moral consideration should not be belittled, as effects upon animal welfare attest. Extension, once again, opens the way for more radical reconsiderations and changes.

Through the use of extension arguments the idea that the interests of nonhumans should not be violated for trivial reasons can be reached. Extension arguments can be a starting point for the process of change in attitudes. Moreover, the basic principle of equality is extended so that, "the interests of every being affected by an action are to be taken into account and given the same weight as the like interests of any other being".[25] The recognition of interests of nonhumans is an important move in advancing moral consideration for them without disregarding differences:

> Giving equal consideration to the interests of two different beings does not mean treating them alike or holding their lives to be of equal value. We recognize that the interests of one being are greater than those of another, and equal consideration will then lead us to sacrifice the being with lesser interests, if one or the other must be sacrificed.[26]

It is apparent that the 'Greater Value Assumption' continues to operate in Singer's principle of equal consideration, but as he emphasizes there is a positive side:

> The more positive side of the principle of equal consideration is this: where interests are equal, they must be given equal weight. So where human and nonhuman animals share an interest – as in the case of the interest in avoiding physical pain – we must give as much weight to violations of the interest of the nonhumans as we do to similar violations of the human's interests.[27]

As an example of how the principle of equal consideration of interests operates in practice, Singer applies it to the human practice of eating sentient beings. The principle does not preclude eating sentient beings, but it does recommend vegetarianism on the grounds of simplicity, if for no other reason. It is simpler to avoid the possibility of violating the interests of sentient beings by eating foods that do not originate from them. A pork roast is one meal for a human, but it is the pig's only life. And to paraphrase the Cynic philosopher Bion, "Boys may throw stones at a frog in jest, but the frog dies in earnest."

So while Singer's principle of equal consideration and 'Argument From Marginal Cases' seem at first sight to furnish strong arguments for greater ethical concern for nonhumans, they turn out to be unduly restricted arguments; a wider ethical reach is needed. The most obvious restriction is that the 'Argument From Marginal Cases' used, applies only to sentient beings. While moral consideration is available to such beings on Singer's arguments, their habitats do not rate consideration. No doubt other principles could be used to bring their habitats into utilitarianism consideration as side-constraints in an expanded utilitarianism. Or they could be brought in as utilitarian arguments about impinging on their happiness by the destruction of their habitat, thus demonstrating the instrumental value of their habitat to them. Or it could be argued, as Passmore does (though again exceeding his own framework), that it is intrinsically wrong to poison the environment of another.

Besides Rodman's objection that extension arguments extend all the problems of the conventional theory with them, a second major objection to extension arguments is presented by Devall and Sessions: "Contemporary humanistic ethical theory is ineradicably anthropocentric, designed specifically to deal with the problems of *human* interaction. When the attempt is made to extend this theory to other animals ..., they are accorded much less moral consideration (less intrinsic worth) than humans".[28] The same is true for extending it to elements of the environment other than animals, thus their objection applies to Leopold's Land Ethic as well as Singer's animal liberation. As implied in their objection, what tends to happen is a rather interesting reversal effect. Take the example of tool use. Tool use was once considered a criterion for defining what set humans apart from the rest of creation and as one characteristic among many once considered to entitle humans to moral consideration. Tool use was considered a sign of intelligence and intelligence was considered a human characteristic worthy of moral consideration. But "because so many primitive creatures display 'tool' use, it is no longer considered a sign of intelligence".[29] Summarily dispensing with tool use as sign of intelligence poses an interesting problem for other morally relevant characteristics: How to treat the link between a characteristic, the significance or relevance of that characteristic, and its use as a defining criterion for moral consideration? One straightforward option is to state that if a particular characteristic has been considered morally relevant and it can be shown or reasonably asserted that that characteristic can be attributed to some species other than humans, then for the sake of consistency that species should be treated with like (but not necessarily identical) consideration due to a human with that characteristic. But there is another option, that where the reversal effect occurs. On this option, if a particular characteristic can be shown to be possessed by or can be reasonably attributed to another species, then instead of according like consideration, the characteristic is dismissed or no longer morally relevant. Plainly special pleading is involved, as the second option becomes untenable only at the point where denial of the moral relevance of a characteristic interferes with interactions between humans. Of course, much else can be abandoned before this point is reached, or theory-saving speciesist ploys can be adopted to maintain distinctions, such as attempting distinctions between mere communication and language.

89

4

A Prominent Deep Environmental Movement: Deep Ecology

Two tenets distinguish deep environmental positions from shallow positions. One is the rejection of the 'Sole Value Assumption', the assumption that humans and human projects are the only items of value. Deep positions share this rejection with intermediate positions. The other is the rejection of the 'Greater Value Assumption', the assumption that humans and human projects always outvalue other considerations and the value of other things. Deep positions are set apart from both shallow and intermediate positions by this latter rejection.

Deep environmental ethics call for a new nonhuman-centred ethic with regard to the environment. It is essential from the beginning to avoid two misconceptions about calls for new environmental ethics: one, they do not replace forms of human to human ethical relationships, and two, they do not dispense with most prevailing forms of ethical concern. On the first misconception, a call for a new environmental ethic is not a call to recast all ethics. Relationships between humans that do not bear on the environment would not be a specific concern of narrow environmental ethic. For example, the respect that spouses do or do not accord each other would not come under the aegis of environmental ethics. On the second misconception, a call for a new environmental ethic does not discard or ignore prudential arguments, instrumental arguments, and extension arguments. In fact many of the considerations introduced earlier, including even those initially advanced for a greening of ethics, have been shallow in character, and may have seemed discordant with proper deeper green concerns. But appropriate use of shallow arguments is not excluded. The critical deeper point is that such shallow arguments are, by themselves, often insufficient. These arguments depend on human interests in the environment remaining the same. These arguments are anthropocentric. Deep environmental positions set anthropocentric concerns within ecocentric concerns. This amounts to an ethical Copernican revolution.

Although humans are regularly said to supply the only known moral agents (except for recognized religious agents, such as God) and, therefore, the only creatures which can observe the ethical restrictions and mandates of an environmental ethic (or any ethic), the ecological community forms the ethical community. Human beings are the moral agents without being ethically privileged in terms of remaining the exclusive objects of ethical concern, or the 'base class' by which ethical concern is measured. In other words, humans are no longer the measure of all things, if they ever were. The rejection of the 'base class assumption', that is the displacement of humans from the centre of the moral universe is disquieting to many. It is disquieting because it is seen as a down-grading of humans or at least of their ethically privileged position. In light of the short-term exploitative position, this means that humans are accountable for their treatment of the environment and things in it, but also, that they can no longer justify by a spurious sense of moral superiority their environmentally destructive conveniences and whims. This is also disquieting to many because it is almost incomprehensible that the value assignments that they have taken for granted for so long could be reassigned to their possible inconvenience or even disadvantage. It might be argued that if humans are shifted out of the centre of the moral universe, then too great an ethical burden is placed on humans for the biotic and ethical community, and they will not bear that burden. But it must be remembered that an environmental ethic or philosophy gets a sure foothold in some quarters, precisely because current treatment of the environment is leading to one disaster after another that threatens human wellbeing and survival. An environmental ethic or philosophy must be viable as well as consistent with environmental and ecological principles. This is exactly what the environmental crisis is forcing Western societies to face – the current treatment of the environment is not viable. Furthermore, for humans to promote their own survival by the destruction of the environment and other species is no more than an imperious delusion. After all, as instrumental and prudential arguments show, humans need other species more than other species need them.

Shifting the frame of reference for ethical consideration from humans to the ecological community also shifts the conceptualization of what is a morally relevant characteristic. Shifting to the ecological community as a frame of reference raises problems of how to respect or value the members of the community. As the late Australian political philosopher, Stanley Benn says:

> What is needed is some body of principles of action that will safeguard valuable things, as presently accepted principles will not. An 'ethic' would then be not so much a systematization of intuitive judgements, but a set of practical prescriptions. Its validity would depend then on the axiological standing of the objects protected by it.[1]

An environmental ethic must be a practical ethic. That is it must be able to translate principles into actions. Furthermore its principles must cover the whole ecological community and not solely humans interests and ethical situations. This then, is a call for a new ethic that runs deeper than instrumental considerations or mere expansions. It is a call for a reconceptualization of ethical relationships and a call to prescribe new normative principles thus incorporating some relationships that have hitherto been dismissed or not considered part of the ethical realm.

One movement that certainly combines practical prescriptions with normative principles is Deep Ecology. Deep Ecology is a deep environmental movement founded in Europe by the Norwegian philosopher, Arne Naess. It has been widely adopted in Australia, as elsewhere, largely through the influence of advocates like Naess himself, who of course coined the term "Deep Ecology" and Bill Devall, a US West Coast doyen of Deep Ecology, both of whom have visited Australia on several occasions. Although many worldwide have eagerly embraced Deep Ecology, it has also attracted many critics. It is primarily an environmental movement and its philosophical core is less than rigorous – intentionally so. The philosophical core of Deep Ecology is called *ecosophy*, a term also coined by Naess, who offers his own favoured version, not compulsory for supporters of Deep Ecology, called *ecosophy T*. *Ecosophy* is a combination of 'ecos' meaning 'household' in the ancient Greek and 'sophia' meaning 'wisdom'. Figuratively, ecosophy is the wisdom of living in harmony with nature, that is one's household in the broadest sense. Ecosophy involves a shift from mere science to wisdom. Devall and George Sessions, another US West Coast advocate of Deep Ecology, have defined ecosophy as "a normative system which includes both norms (or basic values) and factual hypotheses".[2] To blend facts and values is against the former mainstream of Western philosophy, which since the time of Hume adhered to a rigid fact/value distinction. The philosophical core of Deep Ecology, therefore, presents a different ethical footing rather than an extension of dominant systems. It is a different order of ethic rather than a variation on a theme. Deep Ecology is, or was, one elaboration of the position that natural things other than humans have value-in-themselves and that that value sometimes exceeds the value of or had by humans. In other words, Deep Ecology rejects both the 'Sole Value Assumption' and the 'Greater Value Assumption', at least as originally presented it did. But Deep Ecology has been undergoing theoretical changes, as will appear; even its depth is now in doubt.

Deep Ecology is not a science, despite the suggestion in the controversial name. "Deep Ecology is a normative and policy- and lifestyle-oriented theory".[3] Like the science of ecology, which deals with the relationships between organisms and their environments, Deep Ecology is concerned with the place of organisms in their environment, but it is bound up with value judgments about the organisms, their environments, and the relationships between them, and with the wisdom of living by the principles revealed by ecology, and not, as with the science of ecology, with doing experiments that reveal the principles. In a sense, Deep Ecology uses ecology to overcome the embarrassment of science at not being able to convert knowledge into wisdom and state what is most needed – an appreciation of how to value the environment and ecological relationships.

Deep Ecology has four levels (as displayed in Figure 4.1, which adapts Naess). On the first level are the sources of the inspirations, insights, and intuitions of the movement. These may be Christian, philosophical, ecosophical, Buddhist, or from some other source. On the second level is the platform which holds the movement together. The platform consists of principles or departure formulations derived from level one. On the third level are generalized hypotheses. These are generalized ways of behaving towards the environment. The fourth level is the level of actions. These are specific actions in specific cases. If the slogan, 'Think globally; act locally', is applied to these levels, then level three is the global level and level four is the local level. Only at level two is there a consensus, and here the consensus is only a consensus as regards intuitions of the principles and not on their exact formulations. Latitude is left for specific formulations for specific circumstances or bioregions. Thus Deep Ecology is a loosely-knit and open-ended pluralistic movement.

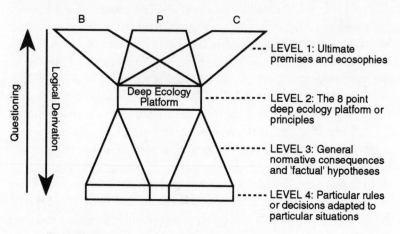

FIGURE 4.1 The double pyramidal structure of deep ecology

NOTE: *In this figure, B, P, and C are not made largely overlapping, chiefly because of the difficulties of formulating agreements and disagreements in relation to texts written in religious language. It is a characteristic feature of Deep Ecological literature that it contains positive reference to a formidable number of authors belonging to different traditions and cultures.*

The departure formulations (sometimes characterized as slogans) on level two form the basic principles of Deep Ecology. Since 1973, when Naess first published them, there have been many formulations and reformulations of the underlying ecophilosophical intuitions fundamental to Deep Ecology. Reformulations of the departure formulations have been described alternatively as a progression and as inconsistent. Perhaps it is best to illustrate departure formulations of Deep Ecology before trying to come to any conclusions about their consistency.

Naess's original 1973 list was:

1. Rejection of the man-in-the-environment image in favour of the relational, total-field image.

2. Biospherical egalitarianism – in principle.

3. Principles of diversity and of symbiosis.

4. Anti-class posture.

5. Fight against pollution and resource depletion.

6. Complexity, not complication.

7. Local autonomy and decentralization.[5]

In 1984 Naess and Sessions published the following 'reformulated' list, an entirely different list, presenting what has since become more or less entrenched as 'the platform':

1. The well-being and flourishing of human and non-human Life on Earth have value in themselves (synonyms: intrinsic value, inherent value). These values are independent of the usefulness of the non-human world for human purposes.

2. Richness and diversity of life forms contribute to the realization of these values and are also values in themselves.

3. Humans have no right to reduce this richness and diversity except to satisfy *vital* needs.

4. The flourishing of human life and cultures is compatible with a substantial decrease of the human population. The flourishing of non-human life requires such a decrease.

5. Present human interference with the non-human world is excessive, and the situation is rapidly worsening.

6.   Policies must therefore be changed. These policies affect basic economic, technological, and ideological structures. The resulting state of affairs will be deeply different from the present.

7.   The ideological change is mainly that of appreciating *life quality* (dwelling in situations of inherent value) rather than adhering to an increasingly higher standard of living. There will be a profound awareness of the difference between big and great.

8.   Those who subscribe to the foregoing points have an obligation directly or indirectly to try to implement the necessary changes.[6]

The diversity of these lists is compounded by other statements of Deep Ecology departure formulations:

Naess 1973	Naess 1983	Naess-Sessions 1984 / Naess 1984	Devall 1979	ROUGH CLASSIFICA - TIONS
—	Intrinsic value (1)	Intrinsic value of life (1)	—	VALUE CORE
Biological egalitarianism (2)	—	—	—	
Diversity / richness (3)	Diversity / richness (2)	Diversity / richness (1)	Diversity (10)	GROUNDS AND BASES
Complexity not complication (6)	—	—	—	
Total field holism (1)	—	—	New person / planet metaphysics (1)	
—	—	—	Objective approach to nature (2)	
—	No negative interference rights, excepting vital needs (3)	No negative interference, etc. (3)	Earth wisdom, limited interference	VALUE AND ACTION COROLLARIES
—	—	—	More leisure (3)	
—	Action obligation (6)	Action obligation (6)	—	

FIGURE 4.2: Survey in note form of main principles of deep ecology from various sources (up to 1985) [continued opposite]

Naess 1973	Naess 1983	Naess-Sessions 1984 / Naess 1984	Devall 1979	ROUGH CLASSIFICA-TIONS
—	Policy adjustments to economic and ideological structures (5)	Policy adjustments, etc. [Also to technological structures] (6)	Interim policy: steady state (15), (8)	
—	—	Objective life quality rather than higher living standard (7)	Life quality rather than quantity of products (6)	
—	—	World population reduction (4)	Reduction of population to optimum (7)	
Anti-pollution/ resource depletion (5)	—	—	Emphasis on pollution and like topics	
Local autonomy / decentraliz-ation (7)	—	—	Local autonomy / decentraliz-ation (11), (14)	
Anti-class posture (4)	—	—	—	
—	—	—	New psychology (3) with rejection of dualisms: man/nature, subject/object, etc.	SURROUND-ING NEW SUBJECTS
—	—	—	New philosophical anthropology (9)	
—	—	—	New objective science (4)	
—	—	—	New education (12)	
Embedding in ecosophy	—	—	Embedded in updated Spinoza (2?)	EMBEDDING PHILOSOPHY

NOTE: *This survey (from Sylvan 1985, 53-4) covers Deep Ecology for the pre-1985 period. With the appearance of Devall and Sessions' text in 1985, Deep Ecology began to change, and to become a wide popular movement. The self-realization directive, largely invisible in the 1970s, began to assume an increasingly important role, in some later sources a dominant role. Therewith*

Deep Ecology became fused with the North American personal growth and improvement movement. Among other significant changes, Deep Ecology belatedly came to include an economic and political reform agenda. Since 1985 the Deep Ecology literature has expanded vastly, and the number of variant formulations of Deep Ecology exploded, excluding any simple listing such as the above survey.

Some significant corollaries flow from this survey. One of relevance, because it means that forms of Deep Ecology may not be deep, is the finding that most Deep Ecology philosophical literature

> fails to present the fundamental value thesis, that intrinsic value is not confined solely to humans or human features. While it can be argued that rejection of the sole and greater value assumptions is implied by what is *said* concerning biospheric egalitarianism (the equal right to live and blossom), the argument is not decisive, since value is only involved indirectly and perhaps only instrumentally.[7]

Some more recent literature has tried to make a virtue of the flight from value principles, berating axiological and ethical approaches to environmental philosophy.[8] Another relevant corollary is that it becomes a daunting task to extract themes of Deep Ecology: both which they are (e.g. is self-realization a directive?) and (as will appear, e.g. with egalitarianism and total field holism) what they mean. It is somewhat tempting to conclude that there is no such well-defined subject as Deep Ecology, just a disparate family of departure proposals. In any case, it has become evident that Deep Ecology is now different things, many different things, to different parties, for some little more than some sort of non-anthropocentrism, for some even less than that. Some authors, Naess is one, do little to dispel the widespread assumption that Deep Ecology covers virtually the whole ecocentric field; for others however, Fox is one, Deep Ecology forms a quite exclusive club, within ecocentrism.[9]

For all the confusion, by drawing upon the various statements, a list of ecophilosophical intuitions can be compiled that illustrates the thinking that sets Deep Ecology apart, in particular from shallow and intermediate positions. Without claiming these intuitions to be the most important or basic, they nonetheless indicate the critical content of Deep

Ecology, as often understood. One such list (compiled by one of us) includes, along with the platform principles (for 1989 n.b.) which do exclude axiological shallowness, the following: biospheric egalitarianism – in principle; relational total-field image; self-realization; diversity and complexity; reduced human interference in the environment involving a fight against pollution and resource depletion; reduction in the human population; bioregionalism; and political and economic reorganization. This list is ordered not simply in terms of how we happened to treat the topics, but primarily in terms of what flows naturally out of the initial (1973) presentation of the Deep Ecology movement.[10].

BIOSPHERIC EGALITARIANISM – IN PRINCIPLE.

'Biospheric egalitarianism – in principle' implies an equal respect for all ways and forms of life. According to Naess's departure formulation, it is "*the equal right to live and blossom ... an intuitively clear and obvious axiom*". It involves, what "the ecological field worker acquires[,] a deep-seated respect, even veneration, for ways and forms of life".[11] Thus the ethical community is rendered co-extensive with the biotic community but the prefix 'bio'- loses it precision and becomes problematic when used in this way. To make matters worse, but to avoid evident objection, biospheric is used in Deep Ecology "in a more comprehensive non-technical way to refer also to what biologists classify as 'non-living'; rivers (watersheds), landscapes, ecosystems".[12] As a result, 'bio'- is thereby used to mean both that which is living and that which is not living; that is, to mean not only what it does mean but also what it does not mean, but is said to mean in Deep Ecology. Similarly 'life' is used in Deep Ecology to mean both what it does mean and what it does not mean, non-life. In part this is a subterfuge, to avoid a telling criticism that the use of 'bio-' and emphasis on 'forms of life' makes the biospherical principle sound simply like a broadening of human chauvinism to bio-chauvinism. This would be no more than the transfer of the putatively privileged status of humans in the ethical sphere to all living beings at the expense of the inanimate.[13] Naess and Sessions concede that the term refers "more accurately to the ecosphere as a whole".[14] The term 'ecospheric' is a delimitation of the class of things in the environment which matter, and that class is said to include *every* natural item. As it is admittedly more accurate, it should be substituted for 'biospheric'. But the substitution brings new problems, which Deep Ecology has not addressed. Namely, how is the ecospheric principle itself now to be formulated? It cannot simply allude to 'ways and forms of life'; it cannot be an 'equal right to *live* and blossom'! So it is equal respect as to what, equal right to do what? (This major unresolved problem will return to haunt the central ecosophical notion of self-realization. How can what is not a higher form of life have a self, strive for its preservation, and *perform* the other feats packed into self-realization? Surely the prominence accorded self-realization introduces inequality, slanting ecosophy in favour of competent forms of life, and importing an ipso-chauvinism.)

The term 'egalitarianism' itself suggests how the items of the ecosphere – or at least of the admitted biosphere – are to be treated: presumably, what is extraordinarily difficult, equally. More plausibly, it denies the privileged status of any species or group in the ecospheric/ethical community. It is with the term 'egalitarian' that the 'equal right' of Naess's slogan 'equal right to live and blossom' comes into play.[15] It is also the term that covertly brings inherent value into play. The term 'right' is not used in a strict philosophical sense. The term is used to imply that it is wrong for humans to interfere with an item in such a way that the item is placed in a dispreferred state or devalued, unless the interference is based on a vital need for humans. What constitutes a vital need "is left deliberately vague to allow for considerable latitude in judgment".[16] Further effects of the criticism that value principles are not adequately represented among the departure formulations can be seen here. The wrongness of interference for non-vital needs is based on the inherent value of the parts or the whole of the ecosphere. Naess and Sessions quote Tom Regan, an American ecophilosopher, approvingly, "The presence of inherent value in a natural object is independent of any awareness, interest, or appreciation of it by any conscious being".[17]

Thus the term egalitarian is intended to imply that the value of an item in the ecosphere is not simply a matter of human need or interest, but a matter of humans recognizing that other beings and items in the ecosphere should be treated with respect because they have value in and of themselves. Egalitarian too does not mean what it seems. For it now must be understood that value is not spread equally throughout the ecosphere. Various items in the ecosphere have different characteristics and these differences require different ethical principles to take account of them and ways of acting appropriately towards them.[18] Ecospheric egalitarianism is modified by the proviso 'in principle', in recognition of differences in required treatment, if not differences in characteristics. Unfortunately for this attempt at repair, the proviso 'in principle' came with a different much narrower interpretation: "The 'in principle' clause is inserted because any realistic praxis necessitates some killing, exploitation and suppression".[19] It is an awkward proviso that sits very uncomfortably with what it restricts: just some suppression in practice of equals? What Naess has gone on to venture reveals that under his developed theory egalitarianism is seriously eroded in principle as well as 'in practice'. For it turns out that not merely do human needs sometimes have priority over non-human needs (the converse is what matters), but that we humans have greater obligations to that which is nearer and dearer to us, normally other parochial humans, and therewith correlative duties and also rights.[20] But those are matters of deontic principle, which infringe the egalitarian principle.

A different attempt at repairing the principle, giving some residual substance to equality, appeals to Singer, who has faced related problems with equality in his theory. Under Singer's discussion of equality as a moral idea, not as an assertion of fact, equality (in principle) contracts into (a near empty) equality of consideration. "It is a prescription of how we should treat humans", or rather consider them, which he extends to other sentient creatures.[21] Deep Ecology simply applies the same moral idea to the ecosphere. There are numerous factual differences among pebbles, kangaroos, humans, and swamps, but ecospheric egalitarianism – in principle would accord them equal consideration, and, in this thin intentional regard, an 'equal respect' and an 'equal entitlement to preservation'. That sort of equality would not however shield them from perhaps devastating practice, since, whatever their actual characteristics, "any realistic praxis necessitates some killing, exploitation, and suppression".[22] Deep Ecology recognizes that use is not prohibited by respect; nor practice by equality of consideration. The trouble with unreinforced equality of consideration, it soon emerges, is ethical emptiness; it is not clear what, if anything material, is thereby prohibited, what practice is excluded. For items *considered* equally (for instance as having some value) can still be treated differentially. Though starting from excessive strength, biospheric egalitarianism has vanished into utter feebleness.

Independently, drastic weakening of the initial principle has been made.[23] Responding to Fox, "Naess confirms Sessions' view that the term *biocentric egalitarianism* is not intended as a formal philosophical position that implies a moral 'ought', but rather simply as a statement of non-anthropocentrism"![24] Some deluded commentators and supporters had imagined that such an elevated principle did more than restate non-anthropocentrism, that it advanced, as it had seemed, positive themes concerning right, values and respect and a certain equality regarding them. Wrong. Now Naess declares "The abstract term 'biospherical egalitarianism in principle' and certain similar terms which I have used, do perhaps more harm than good. They suggest a positive *doctrine*, and that is too much."[25] This unedifying escape is not costless. One serious cost is depth of Deep Ecology. The biospheric principle was the main principle ensuring depth (as we have technically explained it). Without it Deep Ecology may slide into yet another intermediate position.

RELATIONAL, TOTAL-FIELD IMAGE AND HOLISTIC METAPHYSICS.

Once upon a time, a quartz pebble and a swamp had in principle an 'equal right to live and blossom', although in practice a quartz pebble can neither literally nor metaphorically live or blossom. This is not to suggest that the pebble is accorded less consideration or respect because it is considered only a part of the ecospheric whole. The part/whole distinction is not made, according to older content-richer forms of Deep Ecology. By not making, by refusing, a whole/part distinction, Deep Ecology carries with it a "refusal to acknowledge that some life forms [or some ecological items] have greater or lesser intrinsic value than others".[26] Of course this refusal wears a paradoxical appearance; for in refusing to recognize or rank life forms, especially lesser items against wholes made up of lesser parts,[27] parts have already been alluded to, a part/whole distinction presupposed.[28] Nonetheless, this refusal is central

102

to the way Deep Ecologists see and value the environment; and Warwick Fox, Tasmanian-based ecophilosopher, has gone so far as to maintain that such a refusal is the central intuition of Deep Ecology, describing it as "the idea that there is no firm ontological divide in the field of existence".[29] Descartes and other philosophers had maintained that there were firm divides, most notably between creatures with minds such as some humans and the rest of mindless nature. The sweeping Fox dictum rejects not just Cartesianism, but all dualisms (and unless carefully formulated all distinctions).

This refusal can be seen as based on a Gestalt view of existence, as essentially partless. In the relevant English sense 'Gestalt' may be defined as perceiving an organized whole, or complex, to be more than the sum of its parts. (In his distinction between complex and complicated Naess says that the complicated lacks Gestalt; see p.113. The recommended perspective of Gestalt is a "rejection of the man-in-the-environment image in favour of the *relational, total-field image*"[30]. This is taken to mean an integration or identification of humans with non-human nature, indeed of everything with everything else, with no separation or divides, a total holism. Like much else in Deep Ecology, this difference-obliterating holism is never decently explained or rendered coherent. Perhaps the fullest explanation is still Naess's first cryptic statement in his initial presentation of "the Deep Ecology movement":

> a. Rejection of the man-in-environment image in favour of the *relational*, total-field image. Organisms as knots in the field of intrinsic relations [i.e. internal relations]. An intrinsic relation between two things A and B is such that the relation belongs to the definitions or basic constitutions of A and B, so that without the relation A and B are no longer the same things. The total field model [total holism] dissolves not only the man-in-environment concept, but every compact thing-in-milieu concept - except when talking at a superficial or preliminary level of communication.[31]

Deconstructed, unscrambled, this yields a difficult Hegelian doctrine, which taken literally exhibits a familiar incoherence. To escape therefrom the doctrine has to be construed as stating at a superficial level (that still involving 'compact organism') what appears strictly ineffable. It is, moreover, in salient respects environmentally unfriendly, obstructing proper environmental criticism; for environments too are locked up in the total field. For example, the doctrine rejects not only the very way we have explicated non-chauvinistic environmental ethics, beginning with an agent-in-environment notion, but the whole analysis of environmental problems and solutions.

Like all difficult doctrines of Deep Ecology, total holism has been variously diluted or, what is virtually tantamount to platitudinous dilution, abandoned altogether. For instance, such holistic metaphysics disappears not merely from the subsequent platform of Deep Ecology (so it becomes an obscure optional extra for supporters of the movement), but entirely from some of Naess's later expositions of Deep Ecology.[32] Metaphysics is shunted out of Deep Ecology to ecosophy, which is said to 'tie together all life and all nature', a miracle performed through the agency of self. There too Naess moderates total holism, trying "to walk a difficult ridge: To the left we have the ocean of organic mystic views, to the right the abyss of atomic individualism."[33] By contrast, Devall chooses instead to dilute holism, to new triviality, putting the relational, total-field image as "Man is an integral part of nature, not over or apart from nature"[34]. Interestingly, so "re-expressed", the doctrine merges with what Sessions and Naess had collapsed biospheric egalitarianism to, a statement of non-anthropocentrism: "The importance of the intuition is ... its capacity to counteract the ... self congratulatory and lordly attitude [of humans] towards what seems less developed, less complex, less miraculous [mere nature]."[35] Thus the two major doctrines of the original Deep Ecology movement converge, collapsing into a single non-anthropocentric chestnut.

There is nothing new, or even interesting now, in the idea that humans are an integral part of nature, rather than the masters over or foes of nature. Long ago, the Cynics maintained such a doctrine, which condemned the unnaturalness of humans and implied an equality of humans and non-humans. Along the same lines, Spinoza held that humans were an integral part of nature and insisted that humans held no privileged position in nature. Likewise calls for a non-anthropocentric metaphysics also are not new: Spinoza's metaphysics, for one among many, is against anthropocentrism. What perhaps has not been fully realized until recently are the implications of such a metaphysics for ecology. Although this idea too sounds trivial, its impact would not be trivial. The dominant Western paradigm, outlined for instance by Passmore in *Man's Responsibility for Nature*, embodies a subjugation perspective: humans can or should dominate the rest of the environment. Deep Ecology denies the human arrogance that if humans are not yet the masters of nature, then they should endeavour to become so. In this regard Deep Ecology takes an integrated more holistic view. The view of humans as set apart from the rest of the environment is replaced with a view of humans as one species among many and the narrower human-oriented perspective is replaced by a wider bioregional, and ultimately, global perspective.

Human participation is by co-operation with the environment, that is in accordance with the wisdom of ecological principles. The idea of human co-operating with nature is not a call for two rational beings (i.e. treating nature as a rational being and partner in the enterprise) to assist one another in obtaining their mutually desired ends. Rather it is a call to recognize, among other things, that the survival of humans and the survival of other species depends upon compliance with ecological principles and that human dominance over the environment conflicts with compliance with ecological principles. English environmental ethicist Robin Attfield says in his discussion of Passmore's position on human stewardship of nature,

> What needs to be rejected is the attitude and tradition of Despotism, an interpretation of the Biblical belief in man's dominion according to which everything is made for man, nothing else is of any intrinsic value or moral importance, and people may treat nature in any way that they like without inhibition. Instead we should accept that natural processes are not devised or guaranteed to serve humanity, and that manipulating them requires skill and care.[36]

Promoting human interests by a species-centred approach is rejected in favour of promoting human interests by promoting the ecological relationships of which humans and their interests are a part. In other words, the attitude of human domination over or separation from the environment is replaced with an attitude of human interdependence with their environment in which human participation is integrated with the environment. Human ends are still promoted; but they do not

105

dominate, and they are selectively promoted through an integration of humans' needs with the needs of other species and the welfare of the environment, and not to the exclusion of the needs and welfare of the rest of the environment and items within it. An attitude of respect replaces an attitude of despotic domination.

As simple and bland as this resultant doctrine is, it has laid Deep Ecology open to some severe and totally misplaced criticism. In a newspaper article intended to explain and criticize Deep Ecology, a San Francisco journalist, Christopher Reed, talked of the ecological movement containing "hysterical extremism in a manner nastily reminiscent of Hitler's *volk*-ism", thus equating Deep Ecology to a kind of 'environmental Nazism'. Reed quoted Murray Bookchin, founder of the Institute for Social Ecology and well-known opponent of Deep Ecology, approvingly, "Deep ecologists don't place humanity in nature, but under nature, in a crude eco-brutalism that says humans are no more important than snails or ants".[37] This is a blatant misrepresentation of Deep Ecology, but taken together with another departure formulation, the need for a reduction in the size of the human population (to be discussed subsequently) the misrepresentation would become malicious libel, except that it is so ludicrous. Reed claims that Deep Ecologists "want to put nature ahead of man – even if that ultimately means man's extinction".[38] As already explained, exactly the reverse is true. By continually setting human interests above those of the rest of the environment, the ultimate shallow threat is that by destroying the environment humanity will destroy itself along with most other species – a catastrophe Deep Ecology certainly wishes to prevent. While humans and their reckless activities present the single greatest threat to the environment, as emphasized in the discussion of biodiversity, the loss of any valuable species – including humans – constitutes a tragedy devoutly to be avoided.

There is a second aspect devolving from holism worth separate mention. While the term Gestalt is not widely used in Deep Ecological literature any more, the concept has a specific use still, not only to explain the relationship between humans and nature, but to explain value in nature. In traditional Western philosophic debates about value, value has commonly been posited either in the valuer (subjective value) or the object (objective value) or less commonly in the relationship between the valuer and the item valued (relational value). Deep Ecology nows dismisses these traditional views in favour of value Gestalt. The valuer, the item valued, and the relationship between them form a value Gestalt. A value Gestalt overcomes many of the traditional problems about value and drastically shifts away from the idea that humans are the only valuers or the only items that have value. The metaphor of fire can be used to explain value Gestalt. To create fire, oxygen, fuel, and a source of heat great enough to cause combustion of the fuel are required. All three items are needed and none alone is sufficient for fire. Traditionally, value has been explained by picking one of the three elements and attempting to make it suffice for all the elements of value. Value Gestalts bond all three elements of value; all three components together matter. This explanation of value is no longer human-centred, although humans may be the valuers or the item valued in a given value fusion. In a similar (similarly problematic) way, holism of this Gestalt sort breaks down, or rather displaces, the traditional external/internal relation dichotomy, as well as the objective/subjective dichotomy and other dualism. More will be said about this in the discussion of self-realization.

SELF-REALIZATION.

Devall and Sessions consider self-realization and ecocentric egalitarianism the two ultimate norms of Deep Ecology. Despite this, what self-realization is is more intimated than filled out. Naess, for one, is provocatively vague about this central notion of his ecosophy T.[39] The core idea, of unfolding or development of the self, is however both initially self-disclosing and philosophically ancient (reaching back to Aristotle's conception of subject well-being). Self-realization as a value and a goal was strongly emphasized by neo-Hegelians, especially in T.H. Green where it represented subject satisfaction as a whole. But Naess traces the notion historically to Spinoza, construing self-realization as a more active substitute for Spinoza's self-preservation, the basic striving of each competent life form to persevere in its own way of being. Under self-realization the self is to be actively realized, maximally realized. To this old directive, Deep Ecology adds just one significant extension, an expansionary dimension, expansion of a self to a wider Self which may comprehend much more than an individual self (an expansion advertised as enabling ecological wonders). So two points present themselves for infiltration of inegalitarianism and chauvinism into Deep Ecology: in distinguishing within life forms those that have selves, and in discriminating those selves that can expand to Selves. Philosophical explication obviously demands, moreover, a theory of selves, and of Selves, their distinctive features and dynamics, development and realization potentialities. It is supply of this sort of information, presently unavailable, that is commonly cheerfully refused.

Nonetheless some of the (miraculous) roles of self and its realization are indicated relationally in the discussions of other departure principles. Self-realization is said to soften a number of distinctions and assumptions considered almost fundamental to the dominant paradigm. Self-realization makes the internal/external relations distinction less rigid. Self-realization softens the objective/subjective dichotomy. Self-realization is tied in with bioregionalism (though a derived local autonomy! directive) and becoming as one with the land (through the expanded Self). Self-realization is part of dwelling in situations of inherent value and living a life simple in means; rich in ends, in other words quality of life. As to how self-realization accomplishes all those feats, Deep Ecology owes further explanation, yet to be delivered.

"Self-realization individually and at all collective stages presupposes the accessibility of diverse possibilities for personal growth – various activities, and occupations, the utilization of geographical and climatic peculiarities, the exploration of different art forms, broad cultural variation – preservation of diverse cultures".[40] Self-realization is a fulfilment of capacities of the most diverse kinds, as part of the environmental/ecological whole and not as atomic egos. Realizing self, where "'Self' stands for organic wholeness" is, according to Devall and Sessions, the real work of Deep Ecology.[41] Naess has explained some of this work with reference to ecosophy:

> The only contribution of ecosophy to these ... frameworks is the *extension of the concepts* of social self ... to concepts of ecological self. These are *wider* concepts. The social relations of self built up through internal, not external, relations to other *humans*, are complemented with

108

internal relations to non-humans. At the age of two years we already have internal relations to near features of the environment – house, clothing, perhaps some plants or animals. The body of a child may get a new home, but the child itself is not exactly the same. It does not yet belong at the new place.

Only through a rather natural arrogance humans in our kind of culture tend to set humans apart. The slogan 'Not man apart!' belongs to the most potent within the deep ecological movement. One might add 'Not man in environment!' humans are strictly speaking not separable from 'environment' as something 'outside' with which we only have *extrinsic* relations.[42]

By collapsing conventional distinctions and dichotomies, self-realization moves away from, rather than towards egotistical notions of self. By collapsing the internal/external relations distinction, humans are opened to what Naess has called the 'ecological self'. The richness of life, fecundity of reality, and quality of life are explored without reducing experiences to pre-packaged symbols from the shelf or to manifestations of material wealth. As examples of this: it is the difference between learning to play a musical instrument and merely switching on a radio; and it is the difference between playing music as a background noise and actively listening to music. Reflecting on bioregionalism, it is the difference between simply living in a place and dwelling in it by learning about its history, knowing its plants and animals, and participating in what the environment of the region has to offer, rather than simply passing through it or over it. Humans count the environment as part of what they are and their possibilities are shaped with regard to environmental factors, rather than in spite of them.

The breakdown of the internal/external relations distinction is, as Warwick Fox has noted, a recognition of existence as a seamless web with no firm ontological divide. Thus self is realized by seeking possibilities within the ecological context rather than simply within self. Self-realization is environmental self-realization. As the phrase 'ecological self' should make clear, self-realization is not self-aggrandizement of the egotistical sort now practiced. Self-realization is not an environmental glorification and glib justification for an ego-cult. In the words of psychologist, Anna-Marie Taylor "*inner ecology* can't be overlooked as part of the *total* ecology" because "we are all in continuous interchange with the world as *part* of the world" psychologically as well as physically.[43] This means that the distinction between oneself as the subject and all around oneself as the object is also broken down, but not totally obliterated. Identification with the environment and items in it does not imply subject and object are identical. The distinction between identification and identity avoids the crude fallacy: "**a** is related to **b**, e.g. sister of **b**, does not imply that **a** is identical with **b**".[44] With further regard to this breakdown, this also means that the objectification that so often accompanies 'being objective' is discouraged. The individual identifies with her or his environment and treats it as an extension of self and self as an extension of it, rather than separate, unrelated entities.

Self-realization is a normative premise. It is not articulated in the same way for everyone. It may be solitude in the wilderness for some and gambolling in a crowded park for others. Whatever the expression of self-realization its normative value is intrinsic as opposed to instrumental. Self-realization is attempting to recognize the intrinsic value in one's situation as opposed to solely instrumental value.

Against this normative premise it can be argued, as Sylvan does, that self-realization is too psychological, "much too experiential. It renders value a feature of those who experience value".[45] This criticism would imply that any value to be had in dwelling in a situation of inherent value would be value had by the dweller and not shared by the dweller and situation dwelt in. If this is the case, then the entire relational, total-field picture and value Gestalt model are negated. Furthermore, self-realization becomes a matter of maximizing situations of intrinsic value, which in turn means maximizing human-centred values, which reduces self-realization to the same sort of anthropocentric and chauvinistic principle that it is meant to avoid. However, if the internal/external relations dichotomy is dissolved, valuer and valued become as one and some of the force of the criticism is dissipated.

BIO-DIVERSITY AND ECOLOGICAL COMPLEXITY.

This aggregate of principles is essentially aimed at a preservation of environment richness (point 2 of the platform). Diversity and complexity appear separately in various lists of Deep Ecology departure formulations, but they are sometimes connected: "From the ecological standpoint, complexity and symbiosis are conditions for maximizing diversity".[46] In Deep Ecology diversity, complexity, and symbiosis are interconnected. For Deep Ecology diversity, i.e., variety of species as the term is used in Deep Ecology, is the pivotal term among these. Characteristic uses of the other terms can be illuminated in relation to the use of diversity. The interconnections among diversity, complexity and symbiosis may be dually characterized with regard to how they are used in ecology and how they are used in Deep Ecology. A salient feature of Deep Ecology is that it is interested both in how ecology works and how to value it. According to the science of ecology, a direct relationship between diversity and stability does not obtain, although at one time it was thought that it did. It was once thought that diversity led to stability in an ecosystem. In ecological literature stability carries two associated meanings. Stability means either "the ability of the system to remain reasonably similar to itself" or "the system has a greater resistance to changes that are external to the system in their origin".[47] It was therefore important to preserve diversity because the stability, and therefore the existence of an ecosystem may depend on it. Now it appears more likely that diversity may lead to stability in some systems, but not all. That is, diversity increases stability rather than causes it. Even the degree to which the principle of diversity increasing stability works is specific to the ecosystem to which it is applied.

Whether or not diversity increases stability, diversity enhances the opportunities for interaction among species. As diversity increases, the interaction among species becomes more complex. Deep Ecology focuses on this connection. Deep Ecology focuses on an increase in complexity accompanying an increase in diversity, rather than focusing on any connection between diversity and stability. The term 'stability' is not much used in Deep Ecology literature. For Deep Ecology the connection between diversity and complexity is not based on the dubious conventional wisdom that diversity causes stability, but on the wisdom of the reverse claim, "instability increases as a direct function of trophic simplicity".[48] Trophic refers to the nutritional sources available. Trophic simplicity means there are few varieties of food (i.e., fewer species) available as opposed to the quantity of food available (i.e., an abundant monoculture). Trophic simplicity can be understood as relatively few feeding levels in the food chain or few species at particular levels. The wisdom of the Deep Ecology claim is aimed at the idea that a reduction in diversity tends towards instability because it reduces the complexity of interactions, rather than at the idea that an increase in diversity tends towards stability. Ecosystems of comparatively little diversity such as desert regions in Central Australia are stable but delicate. To put this another way, diversity may not cause stability, but simplification reduces stability and increases ecological vulnerability. The specific element of wisdom to be drawn from this is: the simplification of an ecosystem by humans reduces the stability of the system. It is much easier for a pathogen or some other agent to devastate entirely a single species (monoculture), than it is for an agent or even several agents to devastate a diversity of species. For example, a single wheat rust can devastate an entire wheat monoculture. This suggests one reason why diversity is thought inherently desirable.

Symbiosis means 'living together' or 'living together with mutual benefit'. The second use is its more specific and standardly accepted use; the first use is its generic use. Generically the term stands for a number of relationships such as amenalism, antibiosis, commensalism, mutualism, predation, parasitism, and interspecific competition. Without going into detail, the distinct forms of symbiosis described by these terms are:

> Amenalism is an interspecific relationship in which one population is inhibited while the other is unaffected. ... Antibiosis is a specific type of amenalism in which one organism produces a metabolite that is toxic to other organisms. ... Commensalism is an interspecific relationship in which one population is unaffected. ... Mutualism is a relationship in which both populations are enhanced or benefited. ... Predation is a relationship in which one animal species kills another animal for food. ... Parasitism is an interspecific relationship in which one population derives its nutrition from another, usually without killing the host. ... Interspecific competition ... two species cannot occupy precisely the same niche.[49]

About such relationships Deep Ecologists make the point, "the so-called struggle for life; and survival of the fittest, should be interpreted in the sense of ability to coexist in complex relationships, rather than ability to kill, exploit, and suppress".[50] The tendency in the dominant Western paradigm to be in conflict with and to attempt to dominate nature is contrary to most of the interspecific relationships in nature.

The connection between diversity and symbiosis is that the greater the diversity, the greater the opportunities for symbiotic interrelationships. The greater the number of interrelationships, the more complex a system is. For ecology, complexity emphasizes relationships and could be characterized as "a biological community in an area ... interconnected by an intricate web of relationships, a web that also includes the physical environment in which the organisms exist".[51] Complexity is the factor that unifies a diversity of species into a community or system. Complexity is a vague term in Deep Ecology, but Naess makes a distinction between complexity and complication that is not a biological distinction. "The theory of ecosystems contains an important distinction between what is complicated without any Gestalt or unifying principles – we may think of finding our way through a chaotic city – and what is complex".[52] The distinction Naess makes between complicated and complex rests on the unifying interrelationships among the elements of an ecosystem. Complex implies a set of functional relationships whereas complicated does not. It is analogous to the distinction in chemistry between a compound (complex) and a mixture (complicated). In a compound the various elements that make it up are welded together while in a mixture the various elements are simply placed together.

EXCESSIVE HUMAN INTERFERENCE.

What is sought in several of the preceding principles – those of ecospheric egalitarianism, humans as an integral part of the environment, and the plea for environmental richness through diversity and complexity – is "limited interference, human interference to an extent and on a scale *far* below that presently prevailing ... already afforded a basis in the theme of values-in-nature and outside the human sphere, since interference with what is of value is (ipso facto) limited".[53] Despite a number of reservations about its formulation, such a departure formulation or one similar to it is useful in identifying attitudes in the dominant paradigm that need to be subverted or shifted.

Intentional and unintentional human interference with the environment now constitutes an enormous threat on a par with a major glaciation. To supply human wants and needs, profits and whims, vast areas of land are converted from their natural states and diversities to agricultural uses, mines, slums and housing estates, factory floors, waste dumps and other trimmings, such as deserts and carparks. "More than 11 million hectares of forest are destroyed yearly".[54] That is 1250 hectares of forest destroyed every hour. The rate of desertification is 6 million hectares a year.[55]

In the most severe agricultural conversion, diversities are replaced with monocultures intended to exclude all but two species, the cultivated species and the humans who tend it. Yet even less drastic ecosystem modification than monocultural agriculture threatens non-human animal species, because their habitats are modified or altogether eliminated, "By far the greatest cause of a living organism's becoming endangered is removal of its natural 'home' or habitat".[56]

Pollution is primarily a by-product of human intervention in ecological processes and of production processes serving human wants and needs. Forms of pollution, their sources and their effects are well known, but a few examples can serve as a reminder of its pervasiveness. Acid rain is killing forests, poisoning lakes, and damaging coastal marine ecosystems. Chlorofluorocarbons are depleting the ozone layer, while excessive carbon dioxide is helping create a greenhouse effect. The Exxon Valdez oil spill has killed thousands of mammals and birds in Alaska. Residues of manufacturing, runoff of farm chemicals such as pesticides and fertilizers, and eutrophication from high levels of phosphorus compounds and nitrogen compounds (by-products of laundry detergent) pollute water. Repeated applications of pesticides, the dumping of waste products, and the disposal of rubbish pollute the land.

The argument from Deep Ecology is that while humans are entitled to protect their own vital needs, they are not entitled to pollute the habitats of other species for non-vital reasons. In seeking to protect their short-term vital needs they often damage the system that will meet their long-term needs. Long-term viability is part of determining a vital

114

need. The human pollution of the environment is imprudent because it affects humans, but it is also wrong of humans to ignore that they share the environment with other species that have a claim to its use and an interest in its quality. Pollution modifies habitats and directly kills other species. Once again humans need to remember that they need other species more than other species need them. Humans are more likely to miss rainforests than rainforests are likely to miss humans.

Resources depletion is another form of human interference. In an interview, Arne Naess observed, "The shallow ecology movement talks only about resources of mankind, whereas in Deep Ecology we talk about resources for each species".[57] While the most common concern is with the depletion of non-renewable resources – i.e., there are finite quantities of fossil fuels such as coal and petroleum available and once they are depleted they cannot be replaced (not, at least, in human time spans) – there is also a significant and growing concern about what have been considered renewable resources. The depletion of the systems by which renewable resources are produced would be an eco-catastrophe for all species. Supposedly renewable resources are being depleted faster than they can be replaced. A case in point is the circular connection uniting guanays, guano, humans, anchovies, tuna, and plankton. Fifty years or so ago thousands of guanays, a seabird that lives off the coast of Chile on the Chinochos and Sangallan Islands, produced vast quantities of guano which was mined for fertilizer. When chemical fertilizers replaced guano the market for guano disappeared and the Chileans turned from guano mining to anchovy fishing for a livelihood. Anchovies are the principal food of the guanays and of tuna and sea bass, which were also fished by the Chileans. The anchovies fishing was both lucrative and thriving and within a few years the anchovies were fished out. Thousands of guanays, who depended on the anchovies for food, died. With their deaths the guano that once fertilized the sea as well as built up on the islands disappeared. Plankton, the main nutrition for the anchovies, was ferti-lized by the guano that fell into the sea. Without the plankton the anchovies could not prosper; without the anchovies the tuna, sea bass, and guanays could not thrive; without the guanays to produce the guano the plankton could not flourish. Thus in one act of resource depletion the guanays, the anchovies, the tuna, the sea bass, and the humans were all

115

deprived of their livelihoods.

A similar circular connection exists on Christmas Island with Abbott's Booby, nesting trees, and guano mining.

> The conservation of Abbott's Booby provides a classic example of a conflict between the preservation of a scientifically interesting species – which occurs nowhere else and whose value is difficult to assess – and the utilisation of valuable resources. There are surface phosphate deposits derived from bird dropping over countless years. The main breeding areas of the Abbott's Booby coincide with areas rich in phosphate.[58]

The mining of the phosphate destroys the breeding habitat of the Boobys, which are the source of the phosphate. The collection of the resources destroys the source. The source is depleted along with the resource.

Another less complex, yet more tragic example is the depletion of rain forests. 'The lungs of the world' are burnt out of existence. Burning rainforests to create new grazing lands wastes vast biological capital. Living trees are 'carbon prisons' or 'carbon sinks', which remove and keep out of the atmosphere at least some of the 5.5 billion tonnes of anthropogenic carbon emissions per year. One economic functional analysis estimates the value of a single mature tree at $196,250:

> according to Professor T.M. Das of the University of Calcutta … a tree living for 50 years will generate $31,250 worth of oxygen, provide $62,000 worth of air pollution control, control soil erosion and increase soil fertility to the tune of $31,250, recycle $37,500 worth of water and provide a home for animals worth $31,250. This figure does not include the value of the fruits, lumber or beauty derived from trees.[59]

Supporters of Deep Ecology wish to lower the impact of humans on the environment by kerbing human demands on the environment and eliminating waste in the harvesting or using of natural goods and by limiting their use to vital needs.

> [This] argument does not deny human beings the prerogative of habitat modification. To deny them that would be to deny them an entitlement exercised and enjoyed by other creatures, which would be contrary to the integrity and stability of the biotic community. The argument is rather that human beings – in the Western traditions at least – have to

reconceptualize their relationships with nature to develop a heightened sensitivity to what preserves the integrity, stability, and beauty of the biotic community – modification without devastation.[60]

HUMAN OVER-POPULATION.

It is now widely appreciated that "there is a clear association between human population density and faunal destruction".[61] From an ecological perspective, an expanding human population ranks among the greatest of environmental threats, for it is bound up with most major environmental problems. Indeed, an excess of human population such as the world already has, is the source of some of our most serious problems and of many environmental disasters.

Deep Ecologists consider several countermeasures to faunal and environmental destruction, among them reduction of the human population, and adoption and implementation of Deep Ecology's departure formulations, or some other set of principles that give rise to a heightened ecological consciousness. Besides attempting to preserve the environment, another course of action is to reduce the threat of devastation by reducing demands placed on the environment by a rapidly expanding human population. There is an inversely proportional relationship between these countermeasures. The larger the human population the more urgent it is to supersede environmentally insensitive policies with ecologically-inspired and sensitive policies. Conversely, the lower the human population (then – theoretically) the more likely that human devastation of the environment can be contained without the immediate adoption of Deep Ecological principles. Humans have both overrun and overfilled their own niche and the niches of most, if not all, other species. Naess states, "The flourishing of human life and cultures is compatible with a substantial decrease of the human population. The flourishing of non-human life requires such a decrease"[62], and he suggests a maximum human population of one hundred million.[63]

Deeper positions opt for a population sufficient to sustain cultural, economic and other activities, and diversity. They opt for enough humans in certain regions to surpass various arguably adequate thresholds, but not too many more. ... On deeper paradigms there is no commitment to ... population maximisation, or indeed to maximisation at all. All that is sought is ... a large enough human population to provide sufficient variety in significant respects, but not excess, and, generally sufficiency without surfeit. Nor are huge populations required for worthwhile human purposes; often they are inimical to finer human purposes. ... As a homely example of a flourishing culture compatible with a substantially decreased human population we might consider the period of Ludwig Van Beethoven. Beethoven was born in 1770 and died in 1827. The midpoint of his life was about 1800. The world population in 1800 was approximately 900,000,000, about one-fifth what it is today. Beethoven is considered one of the greatest composers in our cultural heritage, yet his brilliance did not stand alone. He was broadly contemporary with Mozart, Hayden, Novalis, Thomas Grey, Weber, David, Goethe, Kleist, Boccherini, Mendelssohn, the Brothers Grimm, Jane Austen, Rossini, Schlegel, Goya, Wordsworth, Coleridge, Donizetti, Hegel, Schelling, Kant, Bentham, Fichte, Thomas Jefferson and Lemarck to name but a few. As this example reveals it is not the number of people, but other factors bound up with quality of life and times, that leads to culture and cultural greatness.[64]

Deep Ecology proposes an option. Without the immediate adoption of some form of Deep Ecological principles, containment of human environmental devastation by reducing and stabilizing the human population is hypothetically possible, but only if accompanied by technological limitation. If a reduced human population retains elements of advanced and environmentally devastating technology, such as nuclear weapons, then a small population would still be able to devastate the world with a nuclear winter or similar catastrophe.

Obviously, despite journalistic innuendos,[65] the call for a reduction of the human population is not a call for genocide, not a call for decimation, not a call for any commission of killing. First, as previously mentioned, the loss of virtually any natural species, including the human species, is to be deplored. If for no other reason, a decrease in diversity is contrary to Deep Ecological principles. For Deep Ecology, there is a core democracy in the ecosphere. Massive reduction in human numbers would very likely enhance diversity in most habitats, because threatened species would have an opportunity to recover and because species with restricted distributions could spread out again. The nonhuman environment cannot sustain, or be expected to continue to sustain, the increasing rates of population growth; and the problems cannot be resolved unless the rate of growth and the population growth cease altogether. Second, no violence is implied. The decrease should be through natural attrition and negative population growth among other things, but not violent methods. No extreme action is suggested reducing human population, no wars, no genocide, nothing macho. Indeed the idea is ludicrous. Deep Ecologists tend to be gentle people, who are opposed to violence; they care about humans, who are part of nature (even if too many of them are trying to set themselves above and apart from it). Although not stated as a principle, non-violence is an implicit norm common to most Deep Ecologists. It is violence against the environment which Deep Ecologists wish to change and they do not wish to perpetuate violence to do it.

A drastic decrease in the human population or even a decline that affected the growth economy mentality would bring about a change in economic and other relationships with the environment. The rapidity and effect of economic changes would depend in part on the time-scale of human population reduction.

BIOREGIONALISM.

The term bioregionalism is not found on any of the lists of main principles. It is, however, used in Deep Ecological literature.[66] The terms used in various lists that cover the idea of bioregionalism are local autonomy and decentralization. The origins of the term bioregionalism are obscure, but inspiration for the concept is drawn from the way that some indigenous peoples, such as Amerindians and Australian Aborigines, related to the places they inhabited. Kirkpatrick Sale, Secretary of the E.F. Schumacher Society, has summed up some elements of bioregionalism, thus: knowing the land, learning the lore, developing the potential, and liberating the self.[67]

Knowing the land involves becoming familiar with the plants and animals with which we share it. On a more sophisticated level, it means learning the potentials of the land to support an integrated and sustainable environmental and economic attitude that includes consideration of the resources for each species, not just humans. For an urban dweller this could include "learning the details of the trade and resource-dependency between city and country and the population limits appropriate to the region".[68]

Learning the lore is discovering the human and natural history of the region. This could include learning how the original inhabitants dwelt there. How they related to the land. It does not mean that it is possible, or desirable, to live like them. It might be compared to following a sporting team. A new fan learns the history of the 'home ground' of his team. The fan learns about past players and the achievements or failures of past seasons. "If nothing else, such history helps us realize that the past was not as bleak and laborious and unhealthy as the high-energy-high-tech proponents try to make out".[69]

Developing the potential follows knowing the place. "Once the place and its possibilities are known, the bioregional task is to see how this potential can best be realized *within* the boundaries of the region, ... constrained ... by the logic of necessity and the principles of ecology".[70] The idea is use but not abuse or squandering the resources of the region, for all users – human and non human.

Liberating the self extends the use of resources to human resources. It is the development of individual potential of the region. Developing the potential of the individual follows from the possibilities opened by centring economic and political opportunities at the regional level and by the individual having a sense of living in place. This will be taken up further in the discussion of self-realization.

In bioregionalism, regions are defined by their natural systems instead of being defined by artificial political boundaries. Political boundaries would be redefined according to bioregions. Bioregionalism is in part founded on the idea of 'living in place'. That is coming to know the place in which you live as "both the source of physical nutrition and as the body of metaphors from which our spirits draw sustenance".[71] This last stipulation about 'spirits drawing sustenance' has the ring of what Reed calls "the unmistakable presence of mysticism, a religiosity based on nature worship".[72] But that is not necessarily the case. It can mean something as innocuous as identifying with one's home town or one's favourite sport team (if the latter is innocuous).

The intention of bioregionalism is to make those peoples living in a bioregion self-regulating. Natural bioregions are self-regulating. This does not mean that each bioregion would be completely self-supporting with regard to economic necessities of its human inhabitants. Cities would not and could not be expected to provide all the foodstuffs necessary to support their populations. As there are different bioregions, there would be different economic zones, but these economic zones would reflect their bioregional characteristics, rather than the other way around as it is now. What it does mean is a radical reconsideration of prevailing political and economic arrangements. Decentralization of power structures and infrastructures and the autonomy assumed at local levels is tantamount to the abandonment of the centralized nation-state.

The call for bioregionalism and the Deep Ecology call for decentralization and local autonomy need not be conjoined, that is, bioregionalism could be adopted within prevailing national boundaries (in many cases) without decentralization of powers and local autonomy. However, as Figure 4.3 indicates, the more likely scenario is the conjunction of bioregionalism with these.

	BIOREGIONAL	INDUSTRIO-SCIENTIFIC
Scale	Region	State
	Community	Nation/World
Economy	Conservation	Exploitation
	Stability	Change/Progress
	Self-sufficiency	World Economy
	Cooperation	Competition
Polity	Decentralization	Centralization
	Complementary	Hierarchy
	Diversity	Uniformity
Society	Symbioses	Polarization
	Evolution	Growth/Violence
	Division	Monoculture[73]

FIGURE 4.3: Simplified Comparison of Bioregional and Industrio-Scientific Paradigms

Chapter 4

ECONOMICS AND POLITICS WITHIN DEEP ECOLOGY

Much has changed recently in deep ecology. In the first wave of deep ecology (roughly that before 1985), while there was a platform claim to the effect that economic policies and structures be changed, there was no due elaboration, and little indication as to what economics and politics would look like under Deep Ecology. With the second wave of Deep Ecology that has changed somewhat. Almost a third of Naess's major text is devoted to economic and politics – within ecosophy.[74]

What appears is, however, disappointing, quite a let down after the build up, earlier in the text, on the gravity of the environmental situation and the need for deep change. In the end there are no clear directives for significant institutional change at all. The conclusion we are reluctantly forced towards is that Naess does imagine, like West Coast Deep Ecologists and despite his own outline of counter-arguments, that requisite change – whatever it is, that too remains unclear – can be effected through individual consciousness raising and personal change, that individual changes will work their way through economic and political structures by democratic means. While however control remains in ecologically alien business and development hands that appears most unlikely.

ECOPOLITICS WITHIN ECOSOPHY

The central political questions for the shallow ... ecological movement are significantly different from those of the deep movement. For the former the task is essentially one of "social engineering", modifying human behaviour ... for the short-term well-being of humans. ... The deep ecological movement sees the present unecological politics as necessary consequences primarily of social and economic priorities, the ways of production and consumption, and only significant changes of this will make the goals of the movement realisable. This implies deep changes of political priorities, and possibly new green parties.

Thus Naess, concluding his remarks on ecopolitics within ecosophy.[75] Only a little reflection shows that this promising account is substantially misleading. Virtually the whole sweep of present political activity, the tasks addressed and so on, is shallow (so far as ethical), not merely that limited re-regulative part that can be accounted 'social engineering', but party political activity (including that of green parties, older or new) and so on. Likewise, but independently, virtually all is short- or (at most) medium-term, though that could change within a shallow setting (e.g. through adjusting to sustainable development). Furthermore, social and economic priorities can be sharply changed within shallow settings; crises and war afford standard examples. For deep ecopolitics then the types of significant changes need to be explained, more *usefully* explained: allusion to unspecified deep changes or new parties is not good enough. In this regard Naess never delivers. "...essentially green politics will be something different."[76] How and what is never explained. "The basic ideals of ecopolitics" alluded to are never explained. There are major obstacles, moreover, in the way of such deep ecopolitics being something different, owing for instance to Naess's commitments to a strong central state, to democratic voting arrangements and mixed capitalism (commitments revealed in his work, partially documented below).

There is *no new political vision* forthcoming from Deep Ecology or ecosophy. Similarly for any accompanying economic vision. Present arrangements are highly incompatible with Deep Ecology, yet no alternatives are really offered. The deep ecopolitics promised is but empty rhetoric, attractive noise signifying naught.

There are real difficulties about what deep ecopolitics within ecosophy *can* amount to, perhaps not insuperable difficulties, but so far unsurmounted ones. It has to be, as already implied, something different from prevailing green politics. Thus Naess cannot simply help himself, as he often does, to green political material, as if this suffices for Deep Ecology and ecosophy.[77] For Deep Ecology is properly included within green, and ecosophy within that again; and the main long ecopolitical frontier is shallow. To try to commandeer it is methodologically unsound. If green politics are to be used, they have to be coupled with Deep Ecology in some other way, such as through a step theory (stepping to what?) or a green pluralism, something Naess does not attempt (though he has the ideological resources to try[78]). Politics more or less as usual in Western Europe, even with a nice green repaint, which is what Naess

123

mostly looks towards and appeals to,[79] cannot satisfy deeper environ-mental aspirations and political purposes.

Deep ecopolitics has to amount to *much* more than individual consciousness raising and change thereupon flowing from individual life-style changes. From the personal development and transpersonal literature on Deep Ecology, and more generally from the West Coast approach to it, it would be easy to gain the impression that such individual practice would suffice. Then, as with imagined religious transformations of societies, a requisite paradigm shift would occur in the human multitude, and all would live happily thereafter. Perhaps through open democratic channels they would elect Deep Ecological leaders – to enact what? Fortunately Naess exposes these sorts of assumptions as wishful thinking; individual life-style changes will not ensure deep-green change, any more than religious change; organization, political action, and more are also needed. For firstly, as Naess notes (appropriately in a section on necessity for power analysis), individual choices even as aggregated, do not determine kinds and quantity of economic production, a point nicely illustrated by 'developments' of Scandinavian rivers for individually undesired power production.[80] What obtains for economic forces also holds as regards political control (which is a main reason why electoral representative democracy has proved no significant threat to the dominant industrial ideology and main power holders). Furthermore Naess follows through with telling remarks about the way in which individuals and whole communities may be locked into damaging systems of practice or production (such as automobile commuting or pesticide application).

> A system of production has imminent forces or implicit aims which mould society. Society accepts the aims as if they were its own and becomes captive to the system [thus to automobile or pesticide use]. Consequently society cannot adopt different aims and values unless the way of production [means of transport or control] is altered. Even when captive we may form ideas about a different system, but these are but expressions of wishful thinking without efforts to alter the dominating system. ... This implies that unless the ideas are acted upon through politics [politics transforming the political system also] there will be no major changes.[81]

Such an analysis not merely supports the important slogan: *fight against depoliticization!* It also means[82] that to be effective ecological consciousness raising has to be directed into political activity. But here any activist encounters two conspicuous deficiencies in Deep Ecology: no explicit political ideology, ideas to guide political action, and no satisfactory action theory, indicating how to put the ideas and evaluations into practice. For action purposes Naess proposes a sort of political pragmatism, with nothing too definite![83] – a disappointing proposal that evades the theoretical issues. Such pragmatism, often unedifying, often unethical, is all too prevalent, and operates for conservative convenience. As regards political ideology, the ecosophical situation is even worse, as will soon appear.

Ecopolitics, like eco-economics in ecosophy and in Deep Ecology more generally, is accordingly disappointing, certainly for those of activist or radical inclinations, keen to change things for the ecosocially better. Although promise of political radicalism is dangled (e.g. "a change of revolutionary depth and size by means of many smaller steps in a radically new direction. ... *The direction is revolutionary, the steps are reformatory.*"[84]), no matching details are supplied. Although some of what must be done, on environmental impact fronts, is clear,[85] how this is to be achieved, what small steps can be managed, is not. For the most part it is, sad to report, politics-as-usual.

While there are now, by contrast with earlier Deep Ecological productions, many remarks on political issues (Naess, like Wittgenstein, tends to offer assemblies of remarks, reminders, fragments, but little sustained argument), many of the remarks offering some new illumination, those on political ideology probe no new depths. What little is presented on basic political issues in particular is evasive, and what is not evasive is superficial. Consider the brief section on "*the big political issues*", within that on "*the basic ideological* choices", and within that again "Capitalism and socialism?"[86]

The evasion on basic issues is not merely offensive for environmental and social reasons, given the enormous and evident effects of capitalism even as restrained and 'mixed' (with a little social justice), but because it sabotages warranted criticism of what is premised on mixed capitalism, dominant economics – criticism Deep Ecology would like to advance and should have made. There is no doubt that in most places now, "capitalist production is the foundation of society; everything else rests upon it"[87]. But it is proving difficult to proceed on changing this entirely undesirable situation; to convert capitalist production into a mere service sector for society. Deep Ecology does nothing to advance the relevant cause, nothing to dismantle capitalism.

Capitalism, one of the most rapacious ideologies that has ever invaded and scourged the Earth, destructive of ecologies and peoples everywhere, itself escapes virtually unscathed, without direct criticism. Naess appears to imagine, like main Christian churches before him, that he can slip quietly by the busy giant ideology, capitalism. Really there never was such an ideology as capitalism, and even if there was something it has been duly criticised by socialism; so let's forget about it. The claim is ludicrous: "While there may be said to be economic policies conveniently called capitalistic, there is hardly any capitalistic political ideology." What, pray, do the encyclopaedias and dictionaries of ideas discuss under 'capitalism', regularly a major entry, if not a political and social ideology; what are the ambassadors and missionaries of capitalism propagating if not the same sort of damaging ideology? Nor is it enough to allude to the "forceful and systematic critique of capitalism" (surely including the political ideology?) from socialism, since these criticisms are launched from a direction different from, and in significant respects incompatible with, Deep Ecology – a direction both shallow and unecological (with socialism typically committed to the increase of all components of impact equations). Deep Ecology should be developing its own critiques, from its own ground, above all of capitalism, but also its false alternatives, such as socialism. One reason why there is effectively no telling criticism of capitalism from visible Deep Ecology, but rather obfuscation and evasion of the issues, is because of commitments to an unspecified variety of mixed capitalism (commitments in part ideologically justified, through norms of self-realization, personal initiative, and following Nature, itself portrayed as competitive).

What is so far tendered on capitalism from visible Deep Ecology is seriously deficient; what is advanced on socialism is marginally better, but hardly satisfactory.[88] No due distinctions are made, for instance, among types of socialism, though certain types, such as centralized and orthodox Marxist forms, are significantly worse news for environments than others, such as pre-Marxist and ecosocialist varieties. The crucial objection to socialism (emphatic about human social organization for social advantage and ends) whatever its form, that all forms tend to be anthropocentric, often disgustingly so, and accordingly that socialism whatever its historic form is incompatible with Deep Ecology, is not made or even approached. Only a new socialism (more a 'communal'ism than a 'social'ism, to adapt one of Naess's hints), a new form yet to be theoretically forged, will cohere with Deep Ecological principles. What criticism Naess does signal is of heavy centralist socialism, fitted out with a stifling bureaucracy. For example, "...socialist slogans still heard [!] are not compatible with" genuine green slogans ": maximize production, centralisation, high energy, high consumption, materialism". Amusingly all such directives also feature prominently in advanced capitalism (e.g. as presently enjoined everywhere in politically orthodox Australia).

The chief criticism made of bureaucracy, correctly taken as integral to modern socialism (though it is also fundamental to advanced capitalism), is that it stifles personal initiative, which Naess takes to be a great good; more generally it suffocates group and community practices, initiatives, decision-making and so on.[89] Ecosophy assumes, along with capitalism, "maximization of personal initiatives will be one of the norms", and accordingly implies, like pristine capitalism, a "hard fight against bureaucratic dimensions". Again this calls for significant qualification: many personal initiatives, like those of anti-ecological developers, need tight reining in, not encouragement to flourish fully, not maximization. Some social structure, perhaps of a quasi-bureaucratic dimension, may be an appropriate way of constraining undesirable initiatives. A main feature of heavy bureaucracy, regulation, "such a minus in capitalism", is also a minus in ecosophy. Mildly critical of socialist greens, Naess is keen to minimize regulation, which he claims can only be achieved "through an internalization of norms", something which "points again to the importance of personal initiatives" (Naess's latent individualism revealing itself again). "The main point here is that we need a change in mentality such that many of the regulations will be unnecessary."[90] But Naess himself has already given the lie to several of these claims: structural change of social and political kinds is frequently required before personal change, even if highly motivated, can be effected. Improved structures can substitute for webs of regulation. Naturally other nonpersonal features also contribute to reducing regulation, above all social co-operation.

Among major political ideologies, Naess mentions four, the further two being anarchism and communism. Admittedly that is better than usual, where a play-off between the false choice between capitalism and socialism – a game now supposed won by capitalism – is widely presumed, communism being equated with socialism, and anarchism dismissed as chaotic background, the state-of-nature relapsed to when political order breaks down. Of communism, Naess has nothing to say, except that it seems less favoured by Deep Ecology supporters than nonviolent anarchism. Such anarchism, though "clearly close to the green direction of the political triangle", Naess too dismisses, too quickly and facilely. "But with the enormous and exponentially increasing human population pressure and war or warlike conditions in many places, it seems inevitable to maintain some *fairly* strong central political institutions." Central states are not, however, deeper investigation reveals, any solutions to wars and warlike conditions, but very often a major part of the problems (responsible for all larger wars and most smaller ones, while stopping but few; major suppliers of military technology and hardware; etc.). Similarly as regards major aggravation of components of environmental impact equations: human population growth, consumerism, damaging technology. Yet it is on this (utterly inverted) sort of basis – the alleged need for strong central authority to safeguard community life "from forces of disruption and violence" – that Naess belittles green utopias that in other places he seems to applaud.[91]

If the periodic stress on strong central institutions – in tension with the emphasis on decentralization elsewhere – is removed, then what Naess does begin to outline of Deep Ecological social arrangements *is anarchistic* in character. In the sketch which is begun (but only begun and abruptly abandoned), a sketch which would constrain (but not fix) ideological options, all the elements are anarchistic: decentralization and local self-determination by communities, self-reliance and mutual aid.[92] No doubt correctly, given present contingent circumstances, Naess sees formidable obstacles, obstructing much "development of [deep] green local communities".[93] There are satisfactory ways around virtually all the sorts of obstacles Naess begins to pile up, ways that should have been recorded, not least for the political use and peace of mind of Deep Ecologists (who should not be left feeling that it is all too hard). Several of the main obstacles listed turn upon the lack of funding and power of local communities as compared with central authorities, towards whom communities have to adopt a submissive and obeisant approach. But central wealth is collected through taxation, confiscation and like methods, from communities themselves and pumped into the centre; power is likewise conceded incrementally and accumulated. Such centralizing economic and political practices are far from impervious to change; monetary pipelines, like others, can be clogged or even cut and diverted; and funds and community power retained locally. Unremarkably changing and deepening social structures involves changing economic organization.

ECONOMICS WITH ECOSOPHY, AND ECONOMIC REORGANIZATION

With economics too, while there is much useful negative material criticizing aspects of prevailing theory and practice, there is little positive material advancing viable alternatives. Much of the negative material, moreover, is now fairly well known (what is not so well known is what to do about the numerous problems revealed, at which there has been much unsuccessful scattered shooting and some wild firing). A put-down would report that most of what Naess has to say against mainstream economics and its welfare variant, his main topics, has been done so often and so well before as to now hold no further interest.[94] For example, Naess chips away at familiar inadequacies of GNP as a measure of progress, or real growth; at the divergence of both demand and need from market demand, with market systems failing to deliver what is needed; at the shortcomings of welfare economics; and at the limitations of shadow pricing. But the negative material does not go nearly as far as might be expected. In particular there is no frontal attack on capitalist market economics, though it is directly (although not solely) implicated

129

in vast environmental degradation. Indeed the treatment is remarkably patchy, often again remarks and notes shunted together, with some surprising lacunae. There is, for instance, little investigation of microeconomics, and nothing much at all directly on markets and their many externalities. What is more, few of the criticisms made within macroeconomics, most of them borrowed from economists, run very deep.

Nonetheless for all the academic familiarity, even dreariness, of Naess's negative material (which may not be familiar to some of his intended readers such as ecologists and potential Deep Ecologists), Naess does occasionally burst forth with some enlivening exasperation (e.g. "The economic mores of industrial countries have ancestors in non-industrial cultures, the mores of husksters"). Moreover Naess does give interesting twists to his retracing of exhausted areas, such as GNP and limits to growth. For instance, sandwiched in his oblique material on 'limits to growth', entertaining fill in otherwise flabby bread, are small harangues to Deep Ecologists to improve their knowledge and their game.[95] There are other surprising passages too where the text converts to an insiders' handbook, where Naess addresses the Deep Ecological movement concerning its tactical approach. For example, supporters need to become informed on economics. Thus Naess proceeds to try to explain GNP, and stock criticisms from renegade economists, for supporters.[96] "It is highly destructive to the Deep Ecology movement for supporters to be silenced because they cannot stand up in discussions with people who are acquainted with economics". Supporters need further to become "competent to take part in economic decision making and take part in informing the public about the consequences of different decisions".[97]

There is, furthermore, good reason for the dearth of positive material advancing a viable alternative. For Naess is tempted to maintain, what is certainly radical, that there cannot be an alternative, a single adequate system: *there is no salvaging of economics*. Briefly, outside a small clear area where objective quantitative measures can be provided and applied, there is reversion to systems of value priorities; it is back to various and rival normative systems, and back to (moral) philosophy.[98] But here, as elsewhere (e.g. as regards ethics), Naess vacillates, in this case, between green reformism and radicalism, between trying to repair fragments of economics or to jettison them. That vacillation is heightened by an accompanying feature: juxtaposed with an underlying hostility to standard economics (endemic in deeper green thought) are intermittent gestures of goodwill and proposals for co-operation.

There are two approaches to goods that cannot be used or applied and are considered defective: repairing or throwing (as well there is indecision, keeping the junk unrepaired). The situation with economics, and environmental challenges to it, resembles that already encountered regarding ethics, as the following table reveals:

	Economics	Ethics
assimilation area made a branch or similar (within dominant disciplinary paradigm)	land, environmental	applied
	typically shallow (perhaps intermediate)	
transformation • reformation • radicalization	ecological	ecological
	usually intermediate	
	usually deep	
rejection	no (notional) subclassificatory title	
	can be any of shallow, intermediate, deep in character	
	commonly partial; e.g. of cost-benefit analysis	e.g. of deontic theory

FIGURE 4.4: Approaches to Unorthodox Theoretical Goods

The dominant approach in Naess towards prevailing economics is revealed in the two final subsections, "The irrelevance of economic growth" and "Misplaced attempts at salvation of GNP", of the section on arguments for ignoring GNP. Fashionable alternatives on offer from reformist approaches, of muted or stationary growth, and of improved

131

GNP measures, are swept aside, the first as irrelevant, the second as leading out of economics, back into philosophy.[99]

Although Naess vacillates in his attitude towards economics, as regards certain parts of public sector economics, his main approach is unambiguous. Along with the GNP measure, cost-benefit analysis is jettisoned; there is no salvaging it generally within ecosophy. Cost-benefit analysis, which is notoriously a shallow utilitarian affair, is rejected for two connected reasons. Firstly, "cost-benefit analysis breaks down in the case of rights", such as rights to unbroken arms and "access to free nature"; instead of acknowledging (ethical) constraints, it is presumed that there are always trade-offs, substitution of items of comparable utility, or dollar value. Secondly, there are many items which cannot be priced meaningfully, such as "irreversible ... *damage* to nature", where "there is no [substitution] relation"; and many concerns which cannot be quantified; in particular, "you cannot slap a price-tag on nature!"[100]

While Naess does not explicitly consider possible rectification, a transformational alternative, through constrained (and deepened) cost-benefit analysis, nonetheless the "boundary or *limitations* strategy" that he mentions, without quite endorsing, amounts to just such a system of constraints. The Norwegian "Master Plan" of protection includes a list of rivers and other natural systems which are declared inviolable, a constraint framework *within* which cost-benefit analyses can be done. As Naess reluctantly recognizes, decision-making is bound to go on within constraint boundaries; "thousands of environmental conflicts in the years to come will not be influenced by the Plan. Therefore the efforts of economists to quantify may still be worthy of discussion" – in a concerted attempt to swing their methods away from unsustainable development towards environmental concerns.

Whereas with GNP growth objectives and cost-benefit analyses, Naess advocates abandonment, with welfare economics he is inclined to urge a sort of replacement, a change to normative systems.[101]

Economics is not what it pretends to be, a unique objective system which delivers absolute scientific results. "Implicitly, economists must take values into account and [include] ethical [components in] solutions to their problems."[102] Values enter directly through such notions as utility, meeting demand, rationality, and development, and indirectly by many other routes. Economists have erected, and are engaged in peddling, one particular normative system among many. That system incorporates at bottom an anthropcentric utilitarianism, nationally partitioned. That is, the particular normative system adopted in mainstream economics is a particularly defective one at that. But that basis is now largely hidden; it is simply presumed. Recently "there has been a dangerous narrowing of the scope of textbooks in economics so that very little of the normative philosophical basis of the field is left, [left visible]. Economics is dried up. We are left with a kind of flat country of factual quantitative considerations".[103] Such quantitative formulations are made possible through presuppositions, such as that entrepreneurs act to maximize profits (and should so act), that consumers possess total market information, and so on, defective presumptions. "We may get a high level methodology, a high level of deduction, a high level of precision, but a certain barrenness from the point of view of norms, barrenness from a point of view of humanity, and extreme danger from the point of view of ecosophy."[104]

Naess makes two attempts to expose particular normative components of economics, and to compare these fragmenting systems with corresponding parts of ecosophy. While neither of these attempts can be regarded as altogether successful, they do make a beginning and do effect some intended damage. The first system concerns standard macroeconomic policy goals. As standardly structured, many hypotheses presumed and required in derivations are suppressed. An example Naess finds contains the unstated presumption that much leisure time is invested in consumption, a presumption which is said to be diametrically opposed to the (dubious) ecosophical hypothesis that "'voluntary simplicity' is ... necessary to achieve much leisure time".[105] In any case, "the postulated relation of derivation between a norm of more leisure time and a norm of more investment *presupposes hypotheses* about how leisure time is to be increased. The fragment is one-sided and reveals according to ecosophy the gigantic illusion that modern industrial society guarantees leisure time."[106] Nor are the basic norms in economic policy placed in a justificatory setting, without which however they appear arbitrary and questionable "*Why should a wise household need high consumption as a basic norm?*"[107] Standard propositions are framework dependent, only holding against a framework of assumptions, which however are ecosophically defective. For example, the standard linkage of employment with economic growth only works against such an assumed background (with present-day truisms about the role of labour and the private sector in economic productivity and growth). There are evidently alternative bases, such as those of ecosophy. "The estimations of economists will therefore only be one set of economic opinions."[108]

A second more extensive exercise looks at welfare economics, which too is revealed as not philosophically neutral, but involving suppressed ideological assumptions and an implicit normative system[109], and shallow questionable norms "which are accepted as basic without justification".[110] In the interests of rapprochement between economics and ecosophy, Naess proposes transforming a fragment of welfare theory into his ecosophical system, by mapping "welfare-theoretic sentences into Self-realisation sentences". In fact the single mapping suggested, of positive choice into improved Self-realization, is not only dubious; it is far from sufficient for a transformation of any worthwhile fragment of welfare theory into Naess's ecosophy system. Rather than pursuing some ecosophical repair for welfare economics, Naess rapidly drops the rapprochement exercise, and proceeds to criticize heavily welfare theory (and indeed economics generally), on the strength of substantially unspecified data "from deep interviews in life-quality research", data presented as much deeper than and superior to results "from market research".[111] And so Naess too quickly concludes "that, whatever the usefulness of welfare theory ... it remains superficial, and it hinders the necessary move from the descriptive to the normative point of view".[112] But welfare theory too can begin with the preference data garnered from these 'deep' interviews; what is shown is merely a recognized gap between those preferences and market-revealed-preferences. No doubt the simple wants market-based paradigm is inadequate and should be abandoned. But it certainly does not disappear without trace into ecosophy T.[113]

There are deep-green economics, plural, a family of them (formulable presumably as normative systems) connected through family resemblance characteristics such as the following: they are against free trade, while not averse to limited fair trade; they are against free (for all) markets, while in favour of certain fair markets; they are against capitalism, and many capitalistic practices, while not averse to certain sorts of personal initiative and enterprise; they are against state socialism, while in favour of social regulation and welfare safety nets; generally, they are against economic activities that violate moral constraints. There is much, of an economic character, that Deep Ecologists should oppose and resist. Little of this is adequately brought out in Naess (or elsewhere in the Deep Ecological literature). Take Trade:

> The economics of the industrial states has tended to favour any increase in trade between nations, and the main thing here is that certain places on the Earth can produce certain products more cheaply and one should always then import from places where they can do these things in the cheapest way, and should export enough to pay for this import. It is very difficult to counteract the force of such argumentation.[114]

Is it? Naess certainly does not proceed to satisfactory diffusion. Firstly, a cheaper price is not a sufficient indicator of value, it does not even indicate superior-value-for-money of a commodity, as other features of production have to be taken into account. For cheaper commodities may be produced through exploitative labour or at heavy costs (through pollution or resource extraction) to environments or to local cultures. Monetary price is not a reliable signal of value, even of economic value more deeply assessed, because of extralities. So we should not always import from where they can do things cheapest, because we should then be fostering environmental degradation or social exploitation or undesirable technology or similar. Secondly, there are regions which – for one reason or another, ranging from incompetence, a laid-back lifestyle and few work ethic commitments, to unexploitative attitudes to local environments and commitments to local ways – have little to export or cannot afford to pay for imports, who cannot or do not desire to trade extensively.[115] Thirdly, there are many other costs to unlimited trade which are not fully costed, but which can show up as extralities, or which are not costed at all. Not fully costed is the oil economy which fuels trade: the costs to future generations of oil supplied too cheaply and profligately, and costs to present creatures in leakage and spills, and so on.

Not costed at all are the virtues of local automony; comparative independence of what happens elsewhere, of the control that may be imposed on local activities; and so on. Naess has sold Deep Ecologists somewhat short on economics too.

An Outline of Deep-Green Theory, By Way of Contrast with Deep Ecology

Many are the possibilities for deep environmental theory; few are those that have been worked out to any extent. One deep environmental theory, different from Deep Ecology, with a distinct Australian flavour is the position developed by Richard Sylvan (formerly Richard Routley) and Val Plumwood (formerly Val Routley), which Richard Sylvan subsequently elaborated. It has come to be called *deep-green theory.*[1] Deep-green theory began with a rejection of human chauvinism, as an expression of deep discontent with the ability of prevailing ethics to take proper account of the sweep of environmental matters, and with an effort to formulate more satisfactory alternatives. Although initiated around the same time as Deep Ecology (in the early 1970s), deep-green theory has since developed in part in opposition to many of the Deep Ecology departure formulations. A main difference between Deep Ecology and deep-green theory is that the former is a movement with philosophical and religious underpinnings, while the latter is emphatically a philosophical approach to environmental problems and issues. Figure 5.1 on the following pages provides an initial comparison of the two positions.

Before taking up some specific points of comparison and difference between Deep Ecology and deep-green theory, a synopsis of deep-green theory is convenient for understanding how deep-green theory combines elements of traditional ethical consideration with a radically new and ecologically-oriented consideration for the environment and parts of it.

Deep-green theory is a deep environmental theory that stands in significant ideological opposition to the dominant technocratic-industrial way. It aims eventually to supply a comprehensive alternative environmental philosophy, a detailed (if not total) philosophical paradigm. Though deep-green accepts the initial 8-point platform of Deep Ecology, the fuller theory is significantly different from Deep Ecology as presently elaborated (e.g. in Naess's ecosophy T), not endorsing extreme

DEEP ECOLOGY	SHARED FEATURES, ETC.	DEEP-GREEN THEORY
	green (action principle, implying commitment to environmental causes)	
	deep (intrinsic value in natural items; rejection of greater value assumption)	
	depth manifested in themes of value distribution	
	and	
biospheric egalitarianism and biocentrism		rejection of class chauvinism; eco-impartiality
	ecological universals (such as richness, diversity, stability, etc.) as defeasible value-making characteristics	
	natural systems as integral and irreducible,	
	coupled with	
extreme holism; total field theme		moderate holism; agent-in-environment acceptable
spiritual alternatives to materialism; salvational as well as individual lifestyle changes	substantial reduction on all prime environmental impact components: human populations; wasteful consumption; damaging technologies	structural changes; social and lifestyle changes
	opposition to dominant social paradigms, and technofix ideology	
bioregionalism	e.g. regionalism and federalism	ecoregionalism
	democratic practices	radical political agenda
	nonviolent practices	anti-nuclear; anti-militarism; social and like defence
	eco-pluralism	radical pluralism; ecological outlook

DEEP ECOLOGY	OPPOSED FEATURES, ETC..	DEEP-GREEN THEORY
looseness, sloganization; departure formulations only; variability	theory and methodology	tight, detailed theory, aiming for precise stable formulation
ambivalence towards ethics		full ethical theory, with axiologic, deontic, and other components
continental philosophy, style, but with hypothetical-deductive methodology		argumentative, analytic and critical approach, with non-standard logic and broad inductive methodology
complicity with pagan and spiritual practices	religious and sceptical commitments	base theory anti-spiritual; complicity with sceptical greened sciences

FIGURE 5.1 (also opposite): A comparison of deep ecology and deep-green theory, in capsule form[2]

holism, biospherical egalitarianism or the ultimate norm (of ecosophy T and transpersonal ecology) of maximizing self-realization. Deep-green theory begins differently, with environmental ethics and removal of human chauvinism, and proceeds differently from Deep Ecology, as the following synopsis reveals .

REMOVAL OF PREJUDICE, ELIMINATION OF CHAUVINISM

In the first place, deep-green theory is, unambiguously, a green *ethical* theory, of deep kind: that is, there is intrinsic value in nature, natural items have intrinsic value, which may predominate over human-based value. In short, the theory rejects both the 'Sole Value Assumption' and the 'Greater Value Assumption'. The theory finds all standard ethics mired in heavy prejudice, a prejudice in favour of things human and against things non-human. The distinctive prejudice concerned is that encapsulated as human chauvinism, which is itself a special case of class chauvinism (for the class, *human*).

Class chauvinism consists in substantially differential treatment, typically discriminatory and inferior treatment, of items outside the class, by sufficient members of the class concerned, for which there is no sufficient justification. Human chauvinism is a feature, a cardinal weakness, of virtually all ethical systems hitherto, so deep-green theory contends. The main argument takes the following form: there is no characteristic, such as those tendered in ethical theories as justifying differential treatment in favour of humans (characteristics such as rationality, language possession, tool making abilities, needs, preferences, sentience, etc.), which first is held by all and only humans, as distinct from not all humans and some non-humans, and which second does justify differential treatment.[3]

The thorough-going rejection of human chauvinism itself sets a program: that of reworking ethics, and indeed much of philosophy, in a way free of chauvinism and humanism, without any specially privileged place for humans. A first part of such a program has already been outlined, that of suitably characterizing *ethics* in a way that does not make essential appeal to humans or their features, or to other favoured groups such as sentients. A further important part of the program involves formulation of ethical theory in a way which avoids entirely group chauvinism. Such a desirable outcome can be achieved simply through formulation of ethical theory in terms of ethically relevant categories, assembled in what is called an *annular theory* (Figure 5.2).[4]

No simple species or subspecies, such as humans or superhumans, no single feature, such as sentience or life, serves as a reference benchmark, a base class, for determining moral relevance and other ethical dimensions. What is required instead is a major shift in perspective, a new focus upon morally relevant categorical distinctions:

> it is not possible to provide criteria which would *justify* distinguishing, in the sharp way standard Western ethics do, between humans and certain nonhuman creatures, and particularly those creatures which have preferences or preferred states. For such criteria appear to depend upon the mistaken assumption that moral respect for other creatures is due

KEY: Notional labels for the interiors of such morally relevant rings (or ellipses), from outer to inner:

Objects of value, objects of ethical concern
Objects having well-being, or welfare
Preference havers, choice makers
Rights holders
Obligation holders, responsibility bearers
Contractual obligation makers

Members of
Homo sapiens

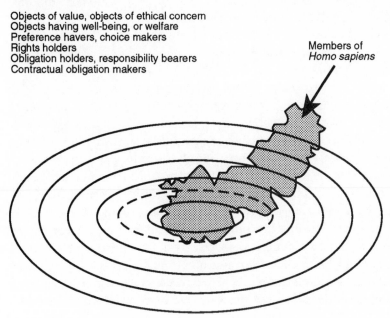

FIGURE 5.2: Annular Picture of Moral Rings in Object Space (and the Position of Humans)

only when they can be shown to measure up to some rather *arbitrarily-determined* and *loaded* tests for membership of a privileged class (essentially an elitist view), instead of upon, say, respect for the preferences of other creatures. Accordingly the sharp moral distinction, commonly accepted in ethics by philosophers and others alike, *between all humans and all other animal species, lacks a satisfactory coherent basis*.[5]

The categorical distinctions proposed by Sylvan and Plumwood do not reject traditional ethical notions such as rationality, self-awareness, having interests; but they do reject the automatic use of the characteristics of one species, or some base class as the model upon which the characteristics of all other species or classes are judged. They also reject

141

as unjustified distinctions that are not "categorical distinctions which tie analytically with ethical notions".[6] For instance, they hold that the human/nonhuman distinction is not ethically significant. Humans enter only in a contingent way into general ethical theory, as holders of these or those features, which features depending upon their respective capabilities and competence.

It is also desirable to have available some more positive principles corresponding to the negative rejection of class chauvinism, principles, so to say, of anti-chauvinism. Principles like the biospherical egalitarianism of Deep Ecology represent an unsatisfactory attempt in this direction. The deep-green improvement on this principle is the powerful *principle of eco-impartiality*, according to which there should be no substantially differential treatment of items outside any favoured class or species of discriminatory sort that lacks sufficient justification. The appropriate impartial treatment is substantially independent of the comparative value of the item treated. Most important, impartial treatment does not entail equal treatment, or equal consideration, and does not require equal intrinsic or other value. Other distinctive principles of deep-green theory will emerge as the ethical theory accumulates.

ON DEEP-GREEN AXIOLOGY

At the core of deep-green theory, however it is formulated, lies a value theory, or equivalent. A fundamental theme and part of what makes the position deep is that a range of environmental items are valuable *in themselves*, directly and irreducibly so, so that their value does not somehow reduce to or emerge from something else, such as features of certain valuers or what counts for them. Value then does not answer back in some way to humans, or sentient creatures (or other value-responsive classes), their interests, uses, preferences, or such like. Many natural items such as forests and rivers, mountains and seashores, are, as it is now often put, intrinsically valuable. They are valuable in their own right – irrespective of whether they are interesting or useful, indeed whether or not any valuers exist.[7] Although value is distributed richly if irregularly throughout nature, and accordingly may be found *in* nature, it is not encapsulated in some separable natural feature, such as life or sentience, or even in defeasible ecological values such as richness, diversity and variety. But of course such ecological features afford criteria for value, and should be represented in methods for trying to assess overall value.

The basic connection is given by the following definitional scheme: an item is good, has positive value, when it has that positive quality by virtue of satisfying relevant criteria, that is value-making factors appropriate to items of that sort. Goodness is one, very general, kind of positive value. Value itself comes in nonquantitative positive, null, and negative kinds (it conforms to a partial ordering under the relation: of no less value than). Though in general the factors that make for value will be broadly naturalistic (such as functioning properly in the case of artefacts like contemporary automobiles, and turning on authentic natural origin and development in the case of life forms), goodness and value are not themselves such naturalistic features, only linked functionally to such features. That is, the 'in virtue of' link, through which the *making* valuable occurs, is a functional one. The upshot, thus far, is a now classic Moorean theory of goodness (as presented in *Principia Ethica*[8]): goodness is a non-natural quality supervenient (by functionality) upon naturalistic features (through ecologically relevant features for good ecological objects). Goodness, and value more generally, do not depend upon the presence of valuers (though possible valuers, hanging out in other worlds, can always be infiltrated); they are not secondary or tertiary or response-dependent properties, and a comparison with shape is superior to one with colour.[9] Furthermore, goodness is essentially indefinable, indefinable outside its circle of own close evaluative acquaintances; so the evaluative/non-evaluative distinction, naturalistic and prescriptive fallacies and other famous Moorean ethico-logical machinery are maintained. The definitional scheme advanced does not afford a definition, because of circularity, and while 'has that positive quality' could be replaced by something else, such as comes out better than indifferent, that does not move outside the evaluative circle. But that is where the resemblance with classic Moorean theory ends: goodness is not an objective quality (nor subjective either, but nonjective); it is not simply intuited (but appreciated through reflective methods based on emotional presentation); no sort of consequentialism is flirted with, not even ideal 'utilitarianism'.[10]

That value notions, though interdefinable, are essentially undefinable syntactically, does not imply that meanings cannot be supplied. They can, by semantical means, through worlds theory.[11] Correspondingly pragmatics can be adjoined, we believe, by coupling context theory. But such semiotical results and investigations do not give the information often erroneously sought under the rubric of meaning, namely epistemic information, about how values are ascertained, knowledge concerning them, how they are acquired or taught and so on. But, contrary to verificationism, semantics and semiotics are one thing, epistemics quite another; and, though there are interrelations between them, they do not collapse to identities. Verificationism, especially in the crude form that meaning is (identically) method of verification or ascertainment, is one of two major Anglo-American philosophical vices. It is matched only by that ubiquitous accompanying vice, reductionism, which has beset and degraded ethics through such forms as naturalism and evolutionism, emotivism and subjectivism. Together they have had an extraordinarily damaging effect on ethics, and encouraged and supported a motley range of defective theories, namely the standard competing positions of set ethical texts.

For epistemics, deep-green theory looks to *enhancement methods*, which, in summary, organize and expand emotional presentation through coherence methods. More intelligibly, the procedure corresponds to a sophisticated coherence updating of the sensory-cum-reflective (or interpretative) accounts of organizing of perceptual information, given by early British empiricists such as Locke. Crudely, competent agents acquire perceptual information through the various senses, which is duly processed, so underwriting simple empirical judgements. To effect the transfer from perception theory to ethics, replace sensory input by emotional presentation, by psychological bases.

Thus the enhancement methodology resembles that sometimes proposed for further acquisition of scientific information. As accumulated empirical information can be further extended through presentational input and assessment for overall coherence, so further evaluations may be arrived at and assessed through a combination of presentational and coherence procedures. At the (marginal) stage where the next round of evaluations is undertaken, the following active ingredients figure (in an idealized breakdown). Firstly, a background stock of judgements and value experience will have been accumulated or inherited. In principle all of this is revisable, and some may be up for reassessment. Apart from the background, which is relevant in an assessment of the overall coherence of the system, what is involved is, secondly, emotional presentation, which corresponds to further perceptional and sense data input; and, thirdly and not independent, coherence processing, which supplies the interpretational and rationalizational components. At bottom, parallelling perception in the case of empirical information, is emotional presentation – gut or visceral reaction in starker forms of acceptance or rejection, but more generally comprehending a variety of sentiments, including overall well-being – and also relational impressions, such as empathy, identification, and so on. As a perceiver perceives shapes, so a valuer *feels* raw value and disvalue. Its basis of perception is sensation, the basis of valuation is emotion. Apprehension of value is seated in emotional, and especially visceral, presentation; but what is apprehended is not to be confused with its apprehension any more than what is perceived. All the warnings about sensation as an information source have to be repeated, with heavy emphasis, in respect of emotional presentation. For example, reliability cannot be guaranteed. Interference with presentation through drugs, alcohol, temporary excitement, or other inputs may render it dubious or unacceptable; or conditioning may have occurred, including substantial cultural conditioning (e.g. so that a person is terrified by harmless spiders but not sickened by bloody massacres of dolphins or seals). As with perception, there are checks on emotional presentation, such as constancy over time and after reflection.

Emotional presentation, supplying primarily inclusions and exclusions or prohibitions, is but the basis of reflective evaluation and value apprehension. The further critical part, coherence processing, builds on the basis taking account of other inputs or controls, including background (which supplies relevant components of already adopted judgements, assimilated subculture, and so forth) and constraints (such as moral substitutional requirements like impartiality, e.g. whether consid-

145

ered judgements hold for substitute valuers, and uniformity, e.g. whether similar acts are judged in similar ways). Essentially, the coherence product consists in asking whether the next or a relevant judgement fits together with what has been accepted, while meeting constraints, without leading to what has been rejected or excluded. If it does fit, it is added to the included side, otherwise it is sent to the excluded side. Because an aim of this rationalization procedure is achievement of some sort of equilibrium, such coherence procedures have gained currency in social theory under the rubric 'reflective equilibrium'. Observe, however, that equilibrium reached at some stage may be lost as new types of problems arise and further information enters. No doubt the whole methodology (like the parallel methodology of an empirically based coherence theory of truth) is highly idealized, and only practicably applicable in rudimentary parts. It does, however, surmount a major theoretical obstacle for environmental value theory: it reveals how in principle a non-reductionistic value theory can function, and deliver a tenable value system. Whether what results, however, is an appropriate deep environmental system will depend above all on the presentational input – the extent to which environmental sensitivity enters and is not suppressed.[12]

BEYOND AXIOLOGY

As it happens, much of deep-green theory has devolved from a specifically environmentally oriented value theory, rather than, say, select Stoic/Judeo-Christian traditions or, more congenially, refined Taoist beginnings.[13] But while deep-green theory does start from value theory, it does not *have* to begin in that way. It could no doubt start from alternative foundations with other primitives, and derive the value theory (e.g. from system well-fare, ethical considerability, etc.). It could develop from a generalized virtue theory investigating certain dispositional characteristics of items and systems (not just virtues and vices of humans of course, as on recently revised virtue theory). Wherever it starts, a fuller ethical theory is desirable, which includes as well as a generous axiological system elaborating deep environmental values and virtues, a deontic framework, supplying obligations, rights, taboos, and similar.

As with the axiological way sketched, so with the fuller ethical theory much changes. For example, deep-green theory advances a semantical theory of deontic notions such as obligation, prohibition and permission, which, by contrast with all standard analyses, takes proper account of moral predicaments and dilemmas, prime moving force in moral change and in reaching ethical maturity. It ventures a different account of rights, which unproblematically allows for animal rights, while not devaluing the important rights currency (because certain rights sometimes claimed for humans lack justification; merely acclaimed rights are not thereby rights).[14] For example, given that a certain wild river is intrinsically valuable, as we can verify by on-site experience and enhancement methods, and given that we duly respect that value, as deontic principles will tell us we should, then we are not free to do as we like with that river, to dam it with concrete, to channel it within concrete, stripping it of its riverine ecosystems. But that does not preclude respectful use of the river, swimming or sailing quietly in its waters, and the like. Value thus guides action, practice, and use. Value is the ground also upon which principles are formulated, principles that are assessed and validated by way of enhancement methods. Important among these are non-interference principles, which exclude unwarranted interference with other preference-havers and unwarranted damage, ill-treatment, or devaluation of items of value. Such deontic principles circumscribe environmentally-limited freedom of action. Given non-interference principles, a major shift in *onus of proof* from homocentric ethics takes place. What is required now is that reasons be given *for* interfering with the environment, rather than reasons for not doing so. Also direct responsibility for environmental interference or modification falls upon those who would seriously interfere or significantly modify, who would tread heavily on the land. Non-interference does not preclude use - only too much use and use of too much. What it does lead to is the theme that, where use occurs, it should be careful and respectful use.

From their rejection of the base class assumption, deep-greens develop a respect thesis founded on three generalizable obligation principles and a no-reduction view of environmental inter-relationships. Briefly stated the obligation principles are: 1) "*not to put others (other preference-havers) into a dispreferred state for no good reason*"; 2) "*not to jeopardise the wellbeing of natural objects or systems without good reason*"; and 3) not to damage or destroy items which "*cannot literally be put into dispreferred states ... but ...can be damaged or destroyed or have their value eroded or impaired".*[15] One or more of these three obligation principles applies to the range of environmental possibilities, e.g. ecosystems, parts of ecosystems, and inter-relationships between the parts or between the parts and the whole of the system.

These generalizable obligation principles are applied in combination with a no-reduction view of environmental inter-relationships. This view maintains a middle ground between 'partist' and holist views. The 'partist', holist, and no-reduction views can apply to the perception of an individual in a social or environmental milieu or to the milieu itself. When applied to the individual, on the 'partist' view the individual is self-contained and disconnected from other individuals. This leads to possessive individualism and egoism. On the holist view the individual is an intersection of holistic elements. This leads to a denial of individual self. When applied to the environment, the 'partist' view would be the equivalent of an atomistic view of the world in which individual items in the environment, such as individual creatures, were the focus of attention. The holist view would focus on entire systems of certain sorts. Observe that such a holism-partism division operates substantially independently of the deep-shallow division, affording an important cross-classification on it.

To take a swamp as an example of the nonshallow situation, on the 'partist' view values, preferences, and other considerations would be focused on the particular mangroves, epiphytes, mussels, barnacles, crabs, water snakes, mud flats, salt domes and other entities and items of the swamp. On the holist view attention is focussed on the swamp as a integrated whole. On the no-reduction view, both the individual members of species and items of the swamp and the swamp as a whole are considered, but the whole is not reducible to the sum of the parts, nor are the parts considered to the exclusion of the relationships and value of the whole. The parts and the whole are respected, but neither is given respect or care at the expense of the other. The middle ground of the no-reduction view provides a locus for what Plumwood and Sylvan term 'the ecological outlook':

> the no-reduction position can provide a suitable metaphysical base for
> an ecological outlook or worldview, in which man is seen as part of a
> natural community, part of natural systems seen as integrated wholes

and with welfare and interests bound up with the whole, and not as, in the typical Western view, a separate, self-contained actor standing outside the system and manipulating it in the pursuit of self-contained interests.[16]

The no-reduction view provides for a policy of respect for the system, its parts, and for the inter-relationships between parts and between the parts and the whole:

> The no-reduction position can thus provide a natural foundation for a genuinely environmental ethic, one which allows human actions to be guided by respect, care and concern for the natural world and rejects the 'Human Egoist' thesis that the only constraints on human action concerning nature arise from the interests of other humans.[17]

Their respect position is not a reverence position. For instance, interference is acceptable, as in the case of "*essential predation*... which is essential to the normal livelihood" of the predator, and where the predator takes for itself no more than it requires for its livelihood.[18] The lives, preferences, choices, and considerations of other species or objects of moral concern are not to be taken as sacred and inviolable, nonetheless the respect thesis requires that good reasons be given for interfering with the environment:

> one starts from a restricted position, a position of no interference and no exploitation, a position at peace with the natural world so to say, and allows interference – not as on Western thinking, restricts interference – for good reasons. The onus of proof is thus entirely inverted: good reasons are required *for* interference, not *to stop* interference.[19]

Under this respect position, the obligation principles are initial restraints on action, rather than inaction. Instead of individuals doing as they wish until they run up against some side-constraint, such as polluting a stream until it affects the lives of humans downstream, by inverting the onus of proof the generalizable obligation principles become constraints on individuals to have a good reason to act before they act. This does not prevent action, but allows for the relationship between the user and the item as well as the preferences, interests or wellbeing, or value of the item to be taken into account. This entails fundamental changes in the present relationship between humans and the environment. One fundamental change is for users to take individual direct responsibility for their impact:

> Respect positions can only be realized in a society in which the basic social structure and economy enables people to take direct responsibility for the impact on the natural world their needs and their satisfaction create.[20]

If human users take direct responsibility for their impact, a reasonable expectation is that they will reduce their impact. Like the supposition behind essential predation (that the predator takes for itself and takes only what it needs), if human users bear the responsibility for producing and managing what they need, then they should produce only what they need and should 'go lighter' on the environment. Direct responsibility would, it is anticipated, lead to:

> a no-waste society in which nothing is produced which does not correspond to genuine needs and in which production is designed to satisfy those needs with a minimum of waste. The possibility of production of material which does not correspond to genuine needs is eliminated by two factors in the economy of cooperative exchange and involvement; first, a direct relationship between the possibility of use of an item and direct expenditure of labour on the part of the user, and second, direct cooperative involvement between producers and users.[21]

When applied to Western cultures, this ethic is intended to face among other issues the economic/ethical conflicts that have been at the core of many of the practical ethical problems concerning humans and non-humans and that have given rise to much of the current debate over the treatment of the environment. The respect thesis is intended, therefore, to reduce the impact of human users on the entire environment. The respect thesis is a strong rival to the Dominion Thesis that has been the ruling and guiding principle of Western thought on ethical relationships between humans and the environment. Like the Dominion Thesis, the respect thesis is a basic set of principles from which a whole range of problems, situations, and questions can be approached. Unlike the Dominion Thesis, it is not the product of centuries of sometimes haphazard, sometimes convenient, sometimes intentional development. It is the deliberate and rational choice to develop a specifically environmental outlook.

AS TO THE WIDER SWEEP OF DEEP-GREEN THEORY, AND ECO-PLURALISM

Deep-green theory thus comprises much more than axiological and ethical theory. It also includes a radical social and political theory, centred upon social anarchistic forms of organization (some glimpsed above), and elements of a radical economic theory, and it is now grounded in a pluralistic metaphysics of transcendental cast.[22] It offers, furthermore, a detailed treatment of such environmentally relevant topics as peace and war, nuclear power, animals and forests, predation and cannibalism, human population and immigration, meanings and purposes of life, ideological sources such as Taoism and Buddhism, and so on.[23]

Some of these further topics flow directly from adoption of a deep-green ethics (some, like ideological sources, are rather *sympatico* positions). There *have* to be changes in social and political arrangements, to improved treatment of other creatures and natural environments, towards less consumptive, populous and wasteful societies, and so on. While a deep-green ethics makes mandatory some such changes, it does not specify which, and it is compatible with a wide plurality of arrangements.

Like authentic Deep Ecology, deep-green theory has taken a significant pluralistic turn. Such a turn has occurred for both theoretical and practical reasons: practically, partly in order to increase constituency (Naess's objective as regards Deep Ecology[24]), partly in order to form as wide a green constituency as feasible. For in these nominally democratic times, human numbers count; with greater numbers supporting green causes there is more hope of extricating humans from the parlous environmental situations many of them now face. With Deep Ecology, a main strategy has been to weaken the membership requirements. It is this that helps account for the remarkable 'adjustment' of the departure themes characterizing Deep Ecology in the decade following the initial presentation (in 1973), in particular the complete disappearance from the platform of the two main initial themes, biospherical egalitarianism and total holism. However there are severe limits as to how far this dilution stratagem, familiar from religious enterprise, can be pushed, while depth (not to mention ecology) is to be retained. For there are numerous greens, especially lax greens, who are not deep, but who are weakly committed to environmental causes for entirely anthropocentric reasons. An obvious alternative, to diluting deep theory, consist in setting deep-green positions within a much wider green coalition, with deep-green thinking as a force, perhaps the vanguard, within that wider amalgamation. This alternative, with deep-green theories designated within a green plurality, is explicitly adopted in deep-green theory, and implicitly adopted in authentic Deep Ecology.[25]

Both these pluralistic turns were also partly motivated by theoretical concerns: at bottom by underlying pluralistic metaphysics. While Naess says that he supports a radical pluralism, intending thereby a pluralism that reaches to metaphysical roots, presumably beneath mere theories, he does not expose these roots. Deep-green theory does; it espouses a thorough-going pluralism which upsets even basic metaphysical and semantical notions, such as actual worlds and truth properties, both of which are rendered plural.[26] For both theories, pluralism affords major advantages, for instance for elaboration of variant theories within a single overarching framework. As a result, only some of apparently favoured political adjuncts to deep-green theory should be regarded as integral to it. Thus, for instance, none of non-violence, pacifism, and organized anarchism are compulsory fare for supporters of deep-green theory. Similarly for such fashionable offerings as sustainable development. By contrast, some commitment to radically changing prevailing political structures to benign environmental ends is essential.

Main obstacles to green coalitions and to targeted action are thereby removed. There does not have to be, does not need to be, any general agreement; a wide diversity of opinion can flourish. That is, adoption of eco-pluralism renders many green tasks and problems much easier. Political organization of diverse green groups to focused green ends is one. As long as an end concerned is acknowledged, combined concerted action to that end can be marshalled from interested groups, irrespective of divergence between groups outside the setting of that end. In this way that major obstacle for radical groups, *ideological correctness* and political and other variants thereupon, can be defeated. Members of groups do not have to have exactly the right positions, and do not need to fritter away long hours attempting to have nuances of a position agreed upon and correctly formulated. Conversely, pluralistic organization can defeat divide-and-dominate tactics from anti-green oppositions, which endeavour to split green coalitions into warring factions, for instance along ideological lines.

FORMALLY RESUMING THE ON-GOING COMPARISON WITH DEEP ECOLOGY

With this much as background, further comparisons and contrasts of deep-green theory with authentic Deep Ecology can be ventured. Both are fully ecocentric, nonanthropocentric theories; that is both are green, both are deep. But while deep-green theory endorses the original 8-point platform of Deep Ecology, taking it indeed to emerge from a fuller characterization of genuinely *deep-green*, it diverges from Deep Ecology on several of its persistent departure formulations, including those by which Naess originally characterized the Deep Ecology movement: namely biospherical egalitarianism and the relational total-field image.[27]

According to deep-green theory, neither of these principles is coherent, as initially formulated; nor, despite Deep Ecological effort, have they obtained satisfactory reformulation. The egalitarian principle yields inconsistent assignments of values, and also of rights, to any ecological whole (e.g. a forest) that is made up of part (e.g. trees, ferns, and so on); the whole and parts have both equal *and* different values (e.g. both the forest has equal value with each tree and also a much *greater* summative value). By contrast, total field integration strictly allows no separation or analysis of the total whole into parts such as individual items at all, no (final) discourse on forests and trees, no presentation of ecological problems![28] As we have seen, deep-green theory moves away both from the extreme holism of Deep Ecology and from the heavy individualistic reductionism of the prevailing technocratic paradigm. Deep-green theory rejects both holistic and partist extremes as inadequate, primarily on metaphysical grounds.[29]

Though rejecting these principles, deep-green theory does offer alternatives, namely impartiality and moderate holism.[30] For instance, rejection of class chauvinism, positively reformulated, yields an impartiality principle: namely, engage in no substantially differential treatment of items outside a chosen (or favoured) class that is discriminatory for which there is not sufficient justification. In short, practice impartiality as between groups. Applied to species, it delivers what was pretentiously dubbed *biospecies impartiality*, so named to correspond to biospherical egalitarianism (it is now better called *eco-impartiality*). These deep-green principles do not however imply any sort of equalitarianism. Only under an insidious (if familiar) confusion does impartiality entail equality, of parts involved. Simple models serve to establish this important point.[31] For example, not taking parts or sides can occur where the sides are not equal; giving every participant its due (or fair handicap) can be accomplished without giving each equal due (where the *same* handicap would be *un*fair!).

Whether the neo-Hegelian directive of maximizing self-realization is a central tenet of Deep Ecology (so Devall and Sessions maintain, as does Fox) or not (Naess tends to make it central to his ecosophy only), it features large in Deep Ecology. But it is a dubious directive for any deeper green position, and accordingly is not endorsed by deep-green theory. The very pedigree of the directive should have alerted suspicion. It emerges direct from the humanistic Enlightenment; it is linked to the modern celebration of the individual human, freed from service to higher demands, and also typically from ecological constraints. Deep Ecologists have attempted to dodge evident problems with the directive, Maximize self-realization!, such as its entanglement with egoism (singleton individual chauvinism) and group chauvinism, by distinctions between types of self-realization (e.g. first between individual-self, small *s* and super-self, capital *S* forms). But, rather patently, this is a subterfuge that does not really succeed, as environmentally-defective super-selves are all too readily invoked (e.g. identification is made with environmentally hostile forces). More generally, the directive (whether maximizing or relaxed to satisizing or other forms), since advancing selves, initiates a sort of self-chauvinism, with selves as a privileged base class. Items of value not incorporated in super-selves are liable to inferior treatment. It is obvious too that many other difficulties loom, such as: what are these selves, and super-selves? What are they like? What has them? What can they do?[32]

An important difference that has not been emphasized so far in the descriptions of the two approaches is the contrast between maximizing and satisizing. The displacement of maximization on deep paradigms by sufficiency, derives in part from a clearer appreciation of limits, environmental limits to growth especially, but limits as well to technology and power, rationality and knowledge (which would seem otherwise to extend environmental limits). The no-limits theme of shallower paradigms, the theme that humans can overcome limits, and always find a way, by science and technology, has a common source with the sole value assumption, a set of prejudices about humans and their abilities, as opposed to other creatures, indeed often a celebration of things human. It is from this illusory Cartesian picture of the unbounded possibilities of humans, with matter wax in their hands, that have developed several fantasies as to the escape from limits; the economic delusion that there are no limits to material growth because substitutes for exhausted resources can always be guaranteed, affordably, through technological means; the grander delusion that there are no limits to human population growth because space is a new frontier, opened again by human ingenuity and technological know-how.

Although deeper paradigms coincide in opposition to this shallower picture, the paradigmatic expansions of Deep Ecology and deep-green theory so far sketched do differ in further crucial respects, including the displacement of maximization. While the damage wrought through maximization of material and economic parameters is increasingly appreciated, and maximization is there resisted and (satisizing) alternatives such as those of sustainability offered, similar restraint is not always shown elsewhere in Deep Ecology. In personal and psychological areas especially, old-style maximization is often persisted with, as with such directives as to maximize self-realization, maximize personal initiative, and so on. Such maximizing directives are liable to be built into expansions of intermediate positions, especially those of a consequentialist cast, which recommend maximization of experience or of interests, or of some other measure of ecospheric utility. The contrast between Deep Ecology and deep-green theory on this point can, perhaps, be illustrated at its sharpest with value. According to deep-green theory, there is no obligation or rational directive to maximize value, still less its putative or degenerate representatives. Yet maximizing value, at least, is the very point which most theories, including Deep Ecology, champion. It is very different as regards maximization of economic impact and performance indicators, such as gross product and throughput, and human population and consumption.

Where deep-green theory and Deep Ecology do agree, apart from full ecocentrism, concerns immediate derivation from platform propositions, such as the reduction of gross human impact, on all main parameters of the consumption input equation. In particular, Deep Ecology and deep-green theory stand together, in broad agreement over the reduction of the human population. Both theories maintain that population policies should be aimed at decreasing the human population and that a substantial but carefully monitored reduction in the human population is compatible with maintaining human values, especially cultural values.

The policy that a substantial decrease in human population is desirable for the conservation of the biosphere is often characterized as extreme. Yet, nowadays such a theme is increasingly widely maintained, by many 'respectable' figures, far from any fringes. Nor need any extreme, threatening or dangerous action result from such a theme. Human population reduction can occur (and is envisaged by Naess to occur) primarily by attrition, by such unthreatening means as a reduction in birth rates, a non-industrial demographic transition brought about for instance through improved education and contraception. If a proper charge of extremism were to be sustained, some coupling to 'extreme' *action*, and adoption or preparedness to adopt such, would have to be established; none has yet been established,[33] and given the ecological state of this world, none now can be.

Beyond ecocentrism, impact factors, and ecopluralism, divergence rather than convergence becomes the predominant pattern. Even so, Deep Ecology and deep-green theory do concur on ecoregionalism, insofar as both recognize a need for political and economic reorganization that takes adequate account of regional environmental concerns and issues, and accordingly depowers present states and also empowers individuals to make real contributions to solving regional environmental problems. Everyone in prevailing exploitative societies is part of environmental problems, and each should have an opportunity to be part of what solutions there are.

While most governments are now elevating environmental concerns on their political agendas, their approaches remain shallow and do not take adequate account of environmental considerations. Most governments are unwilling to proceed far enough on environmental and economic integration. Integration requires that the economy and environment are mutually supportive and sustainable; it requires more than maintaining the environment as a slightly less beleaguered slave of the economy.

Deep Ecology appears to be fragmenting into a variety of different positions. Immediately conspicuous is a bifurcation: a broad division into *social* forms, which are engaged with on-going social issues and which tend to be European, Old-World, in orientation, and *personal* forms, which are concerned with individual consciousness and which tend to be North American, New World. Thus the personal forms are much more concerned with wilderness, biodiversity and similar, and typically committed to the assumption that social change can be effected through aggregated individual change. Still more disconcerting is the appearance both of shallower humanistic forms of Deep Ecology, which abandon the deep value commitments of older authentic forms, and of comparatively contentless forms, which quietly abandon or simply ignore much of the authentic theory, while retaining the fashionable label. In part this proliferation and decadence can be attributed to the weakening of the platform and principles that hold the appropriately pluralistic Deep Ecology movement together. Its founder and main prophet, Arne Naess, has encouraged this degradation by saying that the platform and principles were never more than departure principles, always preliminary, and even open to annual change. As a result, there is little or nothing that stands firm about Deep Ecology; it is not just excessively vague, it ceases to be well-defined. It is one thing, a reasonable thing, for a theory to evolve; it is quite another for a theory to fragment and abandon features that ensure identity through change and all relevant identity criteria.

To glimpse the degradation in Deep Ecology, consider the

diminishing place of ethics and axiology. A very ambivalent attitude is now taken to ethics, while axiology has almost slid off the program (despite the prominent appearance of axiological notions early in the platform). According to deep-green theory, this represents a big mistake. Deep Ecology cannot afford to sacrifice any promising way, especially a way as promising as an environmentally refurbished ethics. By contrast with Deep Ecology, a non-chauvinistic ethical and axiological theory is central to deep-green theory, which aims to include a full ethical component. Though standard ethics may be in doubt because of its shallowness and partiality, ethics is not.

These differences as regards ethics, and more markedly axiology, extend to wider reaches of intellectual endeavour. They extend directly of course to such axiologically-grounded matters as benefit-cost and environmental impact assessments, which according to deep-green theory can be deepened, not inevitably rejected; they extend less directly elsewhere. Very crudely, while deep-green theory falls within a broad Enlightenment tradition (*not* the narrow humanism and the like, sometimes coupled with that tradition[34]), Deep Ecology veers towards an anti-Enlightment opposition.

Deep-green theory is much more heavily committed to analytic and critical methods, to rational procedures, and so on, than Deep Ecology. For example, deep-green theory is less given to the use of slogans and buzzwords, and less into charisma and mystique. It aims to be more precise in both the establishment of terms and the development of the concepts associated with those terms.[35] It remains committed to an enlightened mix of rationalism and empiricism, to a broad naturalism without untestable super-natural elements or spiritual agencies, to the removal and recycling of intellectual and other rubbish (though again it has a different conception of what constitutes rubbish than the dominant paradigm). Unlike Deep Ecology which is keen to welcome almost any sort of movement or bandwagon, under its umbrella, some committed to extensive rubbish, deep-green theory is more discriminating and more critical. It does not refrain from strong criticism even of positions with which it may nonetheless form working coalitions. There is undoubtedly much rubbish in the broad green movement, by no means all of it spiritualistic or religious in cast (though there is much of this), some of it New Age and technowizardry stuff. While forms of spiritualism and technofix are still compatible with deep-green theory (agents with certain commitments to either sort of faith can also adhere to deep-green theory), these are no part of deep-green theory. Nor, though this is not much emphasized and not recognized by some critics (from socialist directions), are they part of the much less exclusive Deep Ecology.[36]

PART II

Action and Educational Directions

Environmental ethics do not stand in resplendent isolation; they relate to environments, and indicate obligatory and recommended practice regarding them. But ethics in splendid isolation, ethics, whatever their calibre, hidden or locked away in ivory towers, will not make any requisite difference. To make a difference they have to be tried out, assumed, lived by; they need experimenters, converts and adherents; they have to be out there, known about and accessible, ready to be taken up, tried, adopted. How all this is to be attempted and accomplished, with a product like ethics, are among the tasks we address. We look, for example, at how to get them out and working in fields and factories, workshops and forests, marketplaces and governments, homes and schools. For these sorts of roles and others, intellectual and academic quality count for much less than a range of other factors, which it is important to distinguish. Some of these factors we soon encounter in investigating how some of the more promising environmental ethics we have outlined in a rough and ready way might be developed into available and more widely used products. Such investigations take us, in this second part, in an unusual and little explored direction from the perspective of environmental ethics.

On the Development of Environmental Ethics

We turn now from critical exposition of the kinds, extent and diverse character of environmental ethics to questions of their development. Development itself has become a rich and ambiguous notion, which carries with it both good and bad connotations; so it is variously a major applauded goal, a dubious or even unintelligible ideal, and a sure route to environmental excesses. When applied to a state or nation, for example, it can imply further industrialization, an increase in GDP, enhanced throughput or pollution – none of them unqualified goods – or else what these are supposed to yield, such as improved economic living standards for certain resident humans. Fortunately such contested notions of development do not make (immediate) sense as applied to an ethic. But many connected notions do. An ethic can be (further) worked out, brought to light, elaborated, complicated, improved, brought to completion, and so on – to enumerate other relevant aspects of development.

FEATURES OF THE PRODUCT AND PRODUCT DEVELOPMENT

The background working model of an environmental ethic as a new product, an economic public good, to be developed and marketed, can be fruitfully applied again. In the development phase of a new sort of product, selection of prototypes is made, and then these are tested and improved, and made attractive to potential consumers. Usually, because of the costs of development, just one or two prototypes are selected. With an intangible good such as an ethic, cost is not such a problem; but time, space, energy, interest and the like remain constraints, indeed apparently such serious constraints that there are *no* really *well developed* environmental ethics. (For that matter there are, astoundingly given the centuries available, few satisfactorily developed standard ethics; utilitarianism would perhaps be the best developed of a poor ethical selection.) There are reasons other than lack of energy, funding support, and so on for this situation, such as the low level of rationality in human affairs, the most important of which here is lack of demand.

Ethics do not count for very much in contemporary predominantly pragmatic times.[1] Of course this situation in turn reveals an incoherence, a deep irrationality, in contemporary practice. Because it is not as if practice proceeds, or could proceed in complex decision making, without value assumptions, and thus core elements of an ethics. But these assumptions are simply accepted under the conventional wisdom, and so remain unexamined and unexposed. A potentially controversial value theory is presupposed and applied, but not acknowledged; an ethic is taken for granted, but ethics are disowned and dismissed. *There* is the incoherence, and the irrationality of an allegedly purely pragmatic approach: where complex decisions are required there is no such pure approach. There are other elements contributing to the erroneous idea that pure pragmatism will work. One is the instrumentalist contribution, according to which all values, or all that are put to work, are merely instrumental, simply means. But the assumed notion of mere means without presupposed ends, like that of instrumental values without presupposed (if concealed) intrinsic values, is likewise incoherent.[2] A similar contribution is made by class-restricted utilitarianisms, such as economism, where underlying intrinsic values are likewise concealed. But they are there, in their full dubiety, underneath the bustle of business, in such shapes as (the uncontestibility of) human want and whim satisfaction, at least for the economically chosen, or else as variations on wants or preferences of select humans.

Typically, people are encouraged to be dissatisfied with the goods, both consumer and others, that they possess and encouraged to want to acquire 'better'. But exactly the opposite normally happens with respect to most ideas and ideosystems, including ethics. Dissatisfaction with prevailing political arrangements, economic organization, religious practices, and the like (by contrast with motorcars, washing machines, and even wives or husbands) is generally heavily discouraged.[3] What we should like to see, what we are aiming to inculcate, is the very reverse: namely, there should be *much less* dissatisfaction with 'goods', so consumption is reduced, and much more dissatisfaction with ideas, especially chauvinistic ethical and economic ideas, thereby opening major pathways to reduced consumption.

In the absence of demand for ethics and new ideological goods, or any but a weak demand, a demand has to be inculcated, *created* as the jargon has it. For weak to zero demand is by no means an unknown phenomenon, especially for substantially new products. (And while ethics is ancient, as are pills and potions, deep environmental ethics is, like genetically engineered pills, substantially *new*, virtually all previous ethics being chauvinistic.) In such cases, product development is coupled to demand generation; for example, features are deliberately built into the product that will arouse interest and stimulate demand. Of course too, products are developed in such a way as to enhance their production features (e.g. scale or mass production) and marketability and advertisability. The issue of marketability raises some interesting and difficult questions, especially for a good like an ethic (and induces significant overlap with the next chapter, on the inculcation of ethics). For it is not well-understood what makes or enables a product to catch on and sell; still less is it known what turns a book into a best seller or a theory into a goer or local winner. An ethic is of course nearer to the difficult end of this rough scale. Certainly some basic threshold features are well enough appreciated; for example, a product must work for a range of standard applications, it must be relatively safe in the short term, it must be easy to understand how it operates, and so on. These sorts of things are enough to make a product a starter, to make sure it is not an immediate failure.

Many such features have to be seen to and taken due care of, in the development of environmental ethics. For example, as regards availability, the ethics must attain a satisfactory and accessible presentation: in grammatical form, not in Farsi and published only in Afghanistan, to indicate some minimal requirements. It will be taken for granted that more minimal features are guaranteed in the essential development process. There are features not far beyond these that are not met by an ethic such as Deep Ecology offers: for instance, a clear and coherent presentation of the doctrine – beyond the eight point platform even this fairly basic desideratum is not satisfactorily met – a consistent doctrine in straightforward areas – but equalitarianism and holism are downright, and rather patently, inconsistent. Such drawbacks should make Deep Ecology, which has however already shown considerable market promise, a poor candidate for development without quite fundamental adjustment (however consider such doctrines as the Trinity, postmodernism, etc.).

Even a well-presented coherent intellectual product may make no market impact. Some reasons for comparative failure are simply due to marketing. For instance, the product was not sufficiently promoted, taken on by influential sponsors (as, luckily for Naess, Deep Ecology was in California), etc. But other reasons may lie with product development. For example, through excessive polishing the product is rendered flat and boring. A product presented through written word, as theories characteristically are, needs entertainment-features (and ideally coupled popular video presentation) if it is to command a sizeable market ; so it will include some sufficient mix of features of *interest*, like the following: some flare, boldness of conception, an illusion of newness even more than originality, limited complexity of basic doctrine, humour, etc. As regards philosophical theories, such as an ethics typically is (though it ventures into the practical also, with deontic recipes), there are further (even more) intangible features, which we can try to collect under the heading: *mystique*.

Mystique is intended to include, when exhibited in high degree, the following types of feature: the impression or illusion of further and great depth; a certain elusiveness, so that matters are never quite pinned down (as with the 'mystique' of the Mona Lisa), but remain tantalizing with more always beckoning; that it will yield more if pressed, and contribute to investigation of a whole range of loosely associated issues From the mystique viewpoint, too much development of a doctrine, toward completion, is a mistake. Much should be merely suggestive, significant features should be left open or mere hints given, and so on. The philosophies that have won most investigation in modern times have considerable mystique, especially those of Kant, Nietzsche, and Wittgenstein. Part of the reason for the neglect of other significant philosophies, like that of the Scottish common-sense philosopher Reid, is their relative lack of mystique.[4] In presenting mystique in this quasi-technical way, paralleling that of the important business intangible *goodwill,* we are sharpening up a little the ordinary sense of 'mystique'. For we have started to explain the type of 'mystery surrounding (relevant) creeds' and doctrines, such as depth, cleverness, range of application, etc. Nor does mystique merely comprise 'professional skill or technique that impresses the layman', though that is part of it: that it gets coverage outside the profession concerned and impresses outsiders including informed lay people; but it also impresses inside professionals, enough of them. But, to be sure, the importance of mystique in intellectual product development may reflect human weakness and show a type of irrationality at work.

One deep environmental position, Deep Ecology again, already enjoys considerable mystique – except that it singularly fails to impress inside professionals (regarding which accordingly something needs to be done). Because of its established mystique, successful proselytization, even evangelical character, it is important to try to hang onto something like it – such as 'Deepened Ecology' – while developing it to remove its evident shortcomings and presentational defects, and to improve its professional image and attractions. Therewith part of the difficult agenda for product development is set: a professionally presentable, marketable, authentic Deep or Deepened Ecology, which also retains however inviting slogans, simple initial doctrine, along with mystique and even a certain tempting obscurity.

Evidently there are available quite a variety of environmental ethic prototypes. Not all of them can be developed given limited time and energy; not all of them are worth developing (e.g. they are unworkable given human limitations or biosocial features; they fail tests, such as that of morality, etc.) Accordingly, we would want to restrict the ethics considered for development, and shall do so: to those whose exposition has already been advanced in the previous chapters. But shallower ethics among those have already had enough exposure, often over a long time period. Nor do they take us environmentally where we want to go, where we need to go, where we ought to go. Following through on these reasons, we naturally select deep environmental prototypes. From deep environmentalism we in fact select two prototypes: a revamped Deep Ecology or Deepened Ecology, and deep-green theory. *How* do we develop these?

Chapter 6

DIMENSIONS OF DEVELOPMENT, AND ATTAINING
DEPTH

Development of an intellectual prototype can occur along several dimensions, including the standard three dimensions: *down*, to further depth and organizational extent, *across*, to increased scope and coverage, including nearby disciplinary fields, and *out*, to application. Such development, well done, can yield a more versatile and comprehensive product. Let us illustrate some of what is desired or required. Almost needless to add, full-scale development, which is an extensive and demanding business is not our present objective.

Despite its claims to depth, Deep Ecology is conspicuously short of depth in important philosophical respects, such as argument, analysis, coherence, structure and stability (even deliberately so). It is in part because of shortcomings ('shortcomings' according to some proponents) in these respects that Deep Ecology has not caught on in (admittedly conservative) academic philosophy. Deep Ecology is conspicuously lacking in arguments for its wider platform; in details of integration of its platform; and in analyses of key notions, for example in explication of its notions of intrinsic value, rights, and so on. These notions are not the same as those of shallower environmentalism (as they do not answer back to class interests, for example), and indeed cannot be the same. So some explanation is in order.

Consider, to illustrate, the notion of *rights*. One of the shapers of Deep Ecology, Devall, proceeds to assert, in a way that does not square well with the general anti-reductionist character of Deep Ecology, that Deep Ecology can dispense with the notion of rights. However not only are several of the departure principles of Deep Ecology formulated in terms of rights, (e.g. "every living creature has a right to live and flourish", "humans have no right to impoverish the environment", etc.), but rights claims provide a powerful way of stating principles, a way that is important and influential, especially in North America, where rights matter and count. They also are important in litigation and provide the basis for much that is relevant in environmental law. So it is not a smart tactic to try to dispense with rights discourse, even if it could be done without residue – which we doubt. Of course while rights remain intact,

166

much can be done and said analytically concerning them: representations can be attempted, analyses offered. But an analysis, which can be given, need not involve a reduction; it need not involve, what cannot be given, any sort of translation scheme.[5] Still the analysis of *right* cannot be quite (if at all) like the standard ones, since, on Deep Ecological percepts, many natural items including plants, insects and much else have rights, in important regards rights every bit as valid as those of humans. Yet Naess is inclined to dismiss any such demands for analysis or explication, and to claim that he is working with the everyday notion of *right*, the notion school-children and others regularly use.[6] That is hardly an adequate response, however, given that school-child usage is predominantly shallow, and that any deeper use will then look like a mere extension. For Naess is supposed to be offering a theory which goes deeper than unsatisfactory moral extensionism – and than school-kid ethics. Deepened Ecology, will accordingly supply, as deep-green theory does, a deep analysis of such notions as *rights*.

In short, Deep Ecology does not supply a proper ethical theory (a defect which some exponents, neglecting the range of implicitly ethical material Deep Ecology helps itself to, have tried to turn into a virtue). While it has a value component and value principles, while Naess makes many suggestions as to value theory, development of a satisfactory axiology, as a prelude to a fuller ethics comprehending the range of ethical notions assumed, is an important outstanding requirement of adequacy.

Such elaboration is but one step in the considerable business of making deep environmental ethics fairly competitive with conventional shallow ethics, and thus removing part of the previous philosophical competitive edge these products enjoyed. Correcting Deep Ecology by requisite argument for its revamped principles is another step in downward development. An example would consist in both negative and positive arguments for the intrinsic value of natural systems, regarded as of merely instrumental value on shallower ideologies. Such arguments would not only back up pronouncements from 'on high', or mere appeal to the upper level of the Deep Ecology pyramid; they could add significantly to the philosophical depth of Deep Ecology.

DEVELOPMENT ACROSS, AND TO BROAD ETHICAL FIELDS

In order to make a deep environmental ethics more, and fully, competitive with prevailing ethics, and indeed in a position to supersede prevailing ethics, deep ethics have to be developed across as well as down, in theoretical scope as well as analytical depth. The dominant Anglo-American ethical theory, which deep environmentalism must be developed to challenge, is utilitarianism; it is that theory which underlies and informs much social theory and politically influential social sciences, economics especially. While utilitarianism has significant rivals on the ethical scene, it is presently without such rivals in the dominant Western theorizing on political economy. But it is precisely there that so much, so damaging to environments, is settled.

Two broad areas relevant to scope can be roughly distinguished, one where utilitarianism does have rivals, one where degenerate forms of utilitarianism, such as parts of economics, reign supreme. The first comprises subject matter still accounted ethics, and often called 'applied ethics', such as medical ethics, bioethics, and different, professional ethics, business ethics, investment ethics. The second includes neighbouring subjects and topics, where *value theory is assumed* (whether acknowledged or alternatively suppressed, in benefits, preferences and the like), such as decision theory, cost-benefit analysis, and so forth. It should be evident enough that deep environmentalism must get into a position to offer alternative (and competitive – equally robust, mathematical, etc.) analyses and to deliver rationally alternative decisions to those that promise, for example, more growth and 'development', paper profit and destruction.[7] In broad outline at least such elaboration is not too difficult.

But why should it be bound to have a role in applied ethical areas? Some examples will make it strikingly clear. Several issues in medical ethics revolve around the production of further humans, several of whom will be highly resource demanding not merely in their production but also in their continued maintenance; other related issues involve the ongoing maintenance of demanding humans. It is not merely that standard ethical approaches assume the dominance of human requirements over all else, and reflect the prevailing celebration of things human; but more,

168

no due account is taken of the environmental impact of these further, often especially support and resource demanding, humans. Again, several issues in bioethics concern the production of items that could well interfere damagingly with natural environments. Much of the work in biogenetics, for example, is geared to the business-as-usual ideology that deep environmentalism considers obsolete and regressive, and accordingly rejects. Types of business ethics are typically set in a similar ideological framework (or at best in a shallow social justice approach). For instance, investment ethics, as now practiced, simply involves avoiding the worst excesses from gung-ho business investment, downright violations of basic human rights and freedoms (thus, e.g. strictures on investment in South Africa, on companies that enslave children, and like conspicuous cases). Deeper investment ethics, by contrast, also systemically avoid investment where environment excesses are involved.

Only a small part of what is required in the way of such development across has so far been undertaken. Much needs to be done in expanding the scope of deep environmentalism. Only in the margins of economic theory have some small but important beginnings been made. For example, deep-green theory offers a rival decision theory, where possibilities are multiplied by environmentally deepened values, not human preferences measures, and the objective is amended from maximization of expected preferences to satisization of expected value.[8] Similar amendments flow through not only into cost-benefit assessment, where environmental costing (expanded cradle-to-grave costing) is expected, but into the main theory of economics, where certainly deep environmentalism does have strong things to say. For, increasingly, environmentalists are appalled by the pronouncements from orthodox economists. (Those like the Ehrlichs, whose pronouncements against economics become more and more strident, afford simple indicators.)

Deep environmentalism rejects the basic assumptions of mainstream (predominantly neo-classical) economics, at both macroeconomic and microeconomic levels. Even Deep Ecology is now fairly explicit about this. The prime macroeconomic objective, of maximizing state economic growth, measured through GDP, is roundly rejected, for familiar reasons.[9] That sort of objective fosters an enormous amount of environmental destruction and degradation. A well-known type of

169

example is that of massive pollution from unregulated production industry followed by an expensive, but partial, clean-up attempt, the total results of which contribute considerably in increasing GDP. That is, 'well-managed' pollution gets highly recommended (partly because double counted!) in national accounting objectives.

Simply increasing GDP can be extremely damaging then, socially as well as environmentally, within a state; and with many environmental excrescences, the damaging products overflow state boundaries. In other ways, too, state economic objectives reach beyond state jurisdiction. Insofar as GDP in one state is inflated at the cost of other states – their environments, their dead forests and lakes, their welfare – its engrossment runs foul of evident ethical desiderata, such as those of equity, impartiality, respect.[10] In place of maximizing economic growth, deep-green theory puts a goal of satisizing upon *selective* and duly *constrained* economic growth. Selectivity is important. For while some sorts of development are no doubt worthwhile, many development projects are definitely undesirable. It is not proposed, then, to freeze economic process into a static state, but to be *much* more selective about the sorts of development undertaken or permitted, encouraged and supported. Constraints are to ensure inter-regional as well as intergenerational equity, selectiveness is to narrow down growth to the kinds of growth that are not environmentally or socially damaging and which are presently required to improve the lot of endangered creatures and seriously-deprived humans. The extent of economic growth required for such objectives is regularly exaggerated; it is well-known that a fraction of current military expenditure (and waste) for 'defence' purposes would suffice.

What is taken as the prime object of microeconomic activity, according to most textbooks, which enthusiastically foster it: maximization of monetary profit, is subject to even heavier strictures. The unfettered profit motive, the pure pursuit of profit (subject only to, what are now however considerable, legal constraints), is again considered undesirable, even sometimes despicable: here Deep Ecology should join forces with what was a widespread sentiment before the rise of capitalism. Pure pursuit of profit by firms and families can, and often does, lead to degradation or desecration of natural environments. It can of course also lead to substantial damage to human beings, destruction of their way of life, through damage to their habitat, ruination of their health, through pollution, and so on. It is surprisingly seldom, however, that these sorts of more extreme but very common states of affairs can be attributed just to profit pursuit. The circumstances where profit can be so obtained have to be arranged; rarely are they simply available. All too often the circumstances are very conveniently supplied by governments or transnational banks, local councils or overseas aid, which provide the requisite, subsidized, infrastructure, or tax havens, or rezoning, that make private creaming feasible. It is typically the alliance of public enterprise with private profit enterprise that wreaks environmental havoc (often too there is corruption, and consumption; there is that famous feedback loop, from private profiteers to public officials). That is not to exonerate gross profit making; it is rather to indicate again that there are other important leverage points in achieving change than through shredding textbook microeconomics. Shredded however it should be; there are known environmentally preferable alternatives. Deep-green theory proposes such satisizing alternatives, in the shape of older refurbished economic notions, like fair margin (for profit) and reasonable rate of return. The refurbishing includes making sure that the returns are environmentally sound, that for instance 'fair' does not include unfairness to other creatures. What replaces profit maximization is then constrained profit satisization, where the thresholds are judged in terms of reasonable returns on gross expenditure of capital, time, labour, ingenuity, and so forth.[11] Such satisization can provide quite sufficient incentive. Of course, as is well-known, profit is not the only motive in firm and corporation operation (though a sufficient, not maximal, level of profit is generally required to stay in long-term business). Several other types of motivation and satisfaction enter; e.g., survival and prestige of the firm, personal opportunities within it, etc. Satisization is correspondingly extended to the broader mix of proper factors.

It is commonly imagined that market operations depend on profit maximization, and that accordingly deep environmentalism is undermining the central institution of neo-classical economic theory, the market, to which all else reduces (so the paradigm would have it). But that is a misconception. Markets, to which much other activity does not reduce, can function and flourish without profit maximization. Satisization is enough. There are in fact, *various* sets of conditions under which supply and demand curves of the right sorts of forms can be attained, equilibrium and market clearance achieved. In particular, fair markets can replace capitalistic 'free' (profiteering) markets. Certainly such markets will require social regulation; but then so increasingly do 'free' markets, to keep them 'free' and competitive, to limit their escalating externalities, and so on.

More generally, fair institutions and arrangements ('fair' in the deep environmental sense) displace 'free' ones; fair trade in place of free trade, fair-for-all in place of free-for-all, fair good in place of free good, fairly-valued good in place of free good, fair whatever-go in place of free whatever-go and of course, fair enterprise in place of free enterprise.[12]

Such a deep transformation of economic theory as here broached, which is intended to extend to practice, inevitably spills over into other institutional areas, especially politics and law. These are in any case shallow and chauvinistic fields, independently in need of deep transformation. But in these fields much less has been attempted than with economic theory, where even so detailed elaboration has not been carried very far. There is much room for development, of a variety of sorts.

There are also many complications. One is the extent of variation of legal and political institutions and accompanying theory from state to state. While economic theory is now a fairly uniform doctrine, imposed almost everywhere throughout the West (thanks to the Americans especially, though much of the basic theory is British), legal and political theory vary much more. Even in those states which operate a precedence-based legal system or a Westminster-style bureaucracy – by no means all states in the West even – there are substantial variations from place to place. The variations also carry advantages; for it is becoming evident that some institutional structures are *much* superior to others, as regards, for instance, obtaining some representation for and standing for the environment and significant environmental objects therein. The West German political system which gives representation to minorities, and the American development of environmental law (deficient though it still is), are striking examples, which perhaps deserve careful emulation elsewhere. In Australia, for instance, environmental law is poorly developed, and generally does not even offer standing (but a first step in satisfactory redress) to damaged environmental systems or their independent representatives.

Some of the very broad changes needed in law and politics are evident from international comparisons together with a small amount of reflection on what could, in theory, easily be achieved. For example, Australian environmental law requires development to achieve, in adequate ways, all of the following: standing, representation, claims and rights, for significant environmental items and systems. Correlatively, penalties for the infringement of those rights need to be introduced, and made adequate and appropriate (e.g. not just fines or fines at all, but restitution and rehabilitation, also embarrassed or ashamed offenders, and so on). To ensure that these matters are taken seriously by company managers and decision-makers (companies are among the main environmental offenders, aided and assisted by governments), limited liability of companies should be substantially rolled back, legal inaccessibility of their directors removed, and so on. In fact there are important social reasons, several intertwined with environmental ones, for major reassessment of limited liability, and the free ride (with expenses paid or avoided), companies have had on communities in which they operate. It is time ethical responsibility was properly sheeted home (to those who syphon off, seldom for worthwhile causes, so much upwardly redistributed monetary wealth). All these considerations and others like them need much more development. But industrial companies are far from the only offenders; farmers collectively are major offenders as regards habitat destruction and severe broad-acre problems such as soil degradation. Agriculture, forestry and fishery too stand in need of significant regulatory restructuring.

It is of course one thing to sketch out in theory what could and should be changed; it is quite another to achieve implementation of any of these sorts of changes, even in perhaps easier institutional areas such as law. Present legal arrangements are heavily weighted in favour of power players such as large companies, to whom access to legal process is also much easier. The whole institutional framework, of which legal institutions are an integral part is substantially under the control of a governing power elite. (Likewise elections serve at best in 'well regulated' representative democracies to shift limited control from one faction of that power elite to another faction. Of course the elite, like all larger human groups, is not uniform but factionalized; but the factions share major ideological commitments, including those to growth, development, wealth, privilege, etc.)

Radical political change, such as deep environmental imperatives require, is still harder to achieve than requisite legal adjustment because it typically calls for major constitutional change – which is extremely hard to achieve short of revolution. But political theory, either reformist or revolutionary, has received little elaboration within deep environmentalism (quite by contrast with Marxism, which provides a useful model), important though it is.[13] There are several, diverse reasons for this, particularly within Deep Ecology: the lack of theory, indeed a certain opposition to too much theory, within Deep Ecology; the considerable American input into Deep Ecology, with its heavy stress on pragmatic individualism; a sense of inevitability (hardly warranted as Foucault shows) about present political arrangements and their resistance to major change – so theory becomes pointless. But on the contrary, theory is very important for guiding change. If the opportunity for revolutionary political change should arise, as it does from time to time, then deep environmentalism needs to be prepared. It should aim for a sound appreciation of what counts as an *appropriate* revolution, and ideally be organized for it, so that control does not go or revert to forces of environmental darkness. Too many revolutionary opportunities have been missed for want of requisite theory and organization. Deep environmentalism is no better organized in this regard than past movements with a major agenda for change, that sought more than token reformation. Moves for political change inevitably link theory to practice, to which we now turn.

OUT TO PRACTICE

A further dimension of development is *out*, to practice. This outward dimension is only as sharply separated from the scope dimension as practice from theory, that is not sharply. To be satisfactory theory has to respond to practice, and rational practice is guided by theory.[14] Deep Ecology does have things to say as regards the relevant practical dimension. It has a descent, not at all well articulated however, down the pyramid (Figure 4.1) from the platform to two levels of practice: a policy level, and a day-to-day and lifestyle level, and it offers a helpful sketch of a Deep Ecological lifestyle.[15] But of course there is two-way interaction, which deep theory should reflect and explain: roughly, theory informs and guides practice, practice confirms and modifies theory.

Deep Ecology insists, as part of its platform, that policies must be changed in most regions. Because

> present human interference with the nonhuman world is excessive and the situation is rapidly worsening (policy item 5) ... policies must therefore be changed. These policies will effect our basic economic, technological and ideological structures. I have not had the courage to go into detail and define what these different structures will be, because we are going to have a lot of different Green societies. We shouldn't have one set of structures imposed.[16]

While a pluralism of policy arrangements and structures is appropriate, matching different Green societies in different regions, and perhaps also, different Green societies, cooperatively sharing regions, this is no excuse for a complete lack of detail. There is, after all, a plurality of Deep Ecological lifestyles also; that does not prevent an outlining of a general framework for such lifestyles, and some details of some. As to policies, there will also be a broad framework. Part of just such a policy framework will derive from the alternative sort of economic theory to which Deep Ecology is now committed. For example, policies will no longer be recommended or selected because of the way they contribute to unfettered economic growth, or to environmentally destructive jobs, or to the flourishing of profit-making markets. Much administration will be decentralized, pushed down at least to bioregional scale. And so on. In consequence, there should arise, in place of prevailing power-hierarchical arrangements, a very different ecologically-founded tiered structure of organization: federal, regional, local. Elements of a radical political agenda lie partially hidden within Deep Ecology, elements that deserve development.

Opportunities for political change at more practical levels, where most grassroots activity takes place, abound (though unsurprisingly much effort fails, as does most standard small business enterprise). An

175

important rather recent initiative in Australia is political campaigning for independent 'green' representatives (whose chances of election are appropriately enhanced in some Australian electoral arrangements through a proportional voting system). If enough of the voting populace can be persuaded – carefully re-educated even – to see the merits of such independent initiatives (which means seeing through extensive propaganda campaigns by the established party machines), some political power can be stripped from the present power elite by entirely constitutional means. Such a move could effectively restore some power to the small people and indirectly to unenfranchised things such as environmental objects, by *depowerment* of prevailing power. But it would be rash so far to expect very much.

Further development is also desirable as regards the connection, or rather lack of it, between deeper environmental ethics and emerging practical politics. Movements such as the European Green Parties and the recently formed green political groupings in Australia and elsewhere, offer political platforms with important environmental implications, but, for the most part, only a shallow environmental philosophy underpins their efforts. The expectation that environmental problems will be decently handled by the present types of governments is nonsense on stilts. At best they have put on Green buttons. Changes are needed by governments, but changes are needed in and within them as well. Furthermore, it is obvious from the rise of these movements that although environmental awareness is growing – its growth is slow and often stunted. On the theory side greater variety, depth, scope and detail is needed in developing and translating environmental ethics into political action. On the practical side better practice is needed in getting the theory down not merely, or even so much, to the people at the grassroots but, more important, under present hierarchical arrangements, up to the people and institutions at the top. Which brings us to the business of implementation, after which issues concerning practice will be resumed.

On Ways and Means of Marketing, Propagating, Inculcating and Implementing Environmental Ethics

After product development, planned during development, lies product marketing, winning product acceptability, generating increased product demand. A main method deployed, especially (but not only) in the private section of business enterprise, is that of repeated advertising of a product. This is a main, and often powerful, way of announcing a product, of proclaiming its desirability and other alleged advantages. Repeated advertising is one method of inculcating something, that is, of impressing something (mentally) by emphasis or repeated repetition, of enforcing or instilling something. But many other means of communication offer inculcation procedures, especially when the message can be repeated frequently. All the main media channels enable inculcation procedures, as does instruction of a range of kinds.

MARKETING STRATEGIES FOR IDEOLOGICAL PRODUCTS

The procedures of product marketing are directed primarily at product sale. When the product has been sold the main work of marketing has been done – even if nowadays there are increasing after-sales commitments, to meet guarantees, to ensure product reliability and safety (and so on, environmental acceptability), and to foster further sales. With a product like an ethic, there is no comparable market highpoint, no clinching a sale, most likely no conversion either. Though ethics remain unusual goods, not themselves admitting of prominent shelf display, nonetheless much of the majority of marketing theory and the art of its practice can be brought to bear, as with newer intangible computer products. Though the medium by which the ethic is presented may have to be purchased (the book or tape or seminar), an ethic, like ideas generally, is still a relatively 'free' good (not so far patented or owned, if even named).

Like ideas and other intangibles too, so with an ethic that is to be applied, after-marketing activity is all important. The ethic is like a seed; it has to fall upon ground that is not too barren, in a region that is not too unfavourable to grow at all, and to really flourish much more still is required. If a purchaser takes or purchases an ethic, in an ethical text say, stores it away and does nothing with it, then it has fallen on so far barren ground. As with a gift of a language package, the gift is only really received to intended advantage if the package is put to work, and the receiver ends up with increased capabilities or fluency in the language concerned. For success with an ethic, more than marketing is required, more than inculcation as with a foreign language is required too: *acting upon* the ethic is required. Absorption and implementation is what is sought. We, as suppliers, want people, institutions and organizations, to take a deep ethic, imbibe enough of it, and *act* according to it.

To try to inculcate a whole ethic, like inculcating a creed or ideology, would be an ambitious undertaking, a task unlikely to succeed with regard to the mass of the older population. (It would be like switching from imperial to metric measures, but with difficulties multiplied up many times.) The most that could reasonably be expected on a broad human front is that some of the main messages of a environmental ethic got across, perhaps in mnemonic slogan form (so we should write versions of them that way, as Deep Ecology does). And then we could hope that these were duly endorsed and acted upon. For mere inculcation, though perhaps meritorious (depending on the merit of the doctrine), is no guarantee that what is inculcated is accepted or that it will be acted upon appropriately. No end of children are still obliged to learn antiquated and sometimes pernicious creeds or ideologies, so that for example they can recite them word perfect. That, fortunately, is no insurance of full acceptance, though too often the messages are imprinted and stick.

There are other important reasons for a much more piecemeal and fragmentary approach than inculcation of a whole ethic. One is the question of time, the urgency of many environmental problems, as compared with the time it commonly takes (a generation at least, assuming success) to inculcate an ethic. As far as getting across to people who can make a difference (including ultimately enough voters, in places where there is a genuine choice of candidates), powerful messages and images conveying simple critical elements of a deep ethic are a better investment and pay off more rapidly. But of course the detailed critiques of prevailing assumptions, and the elucidation of alternatives, have to be there somewhere also, to make for intellectual respectability, to provide justification, and offer fallback in arguments, and so on. That is why proper development is important, that is where, in any case, the powerful messages will largely derive from; ideally, they will condense elements of worked out theory.

As with developing environmental ethics, so with promoting and marketing them, it is worth stressing at the outset that not all environmental ethics are created equal. To promote only shallow environmental ethics is hardly good enough, for reasons that are already apparent and that will be reinforced later. Shallow ethics are for the most part all right in a limited human setting as far as they go, but they do not go far enough. It is deeper environmental ethics that should be developed and promoted. It is a *substantial* change that is wanted. It is not just the stopping of impending environmental disasters to humans that is required, but an appreciation of the intrinsic value of other things that share the environment with humans that is needed. Promoting only shallow environmentalism could leave the world devoid of its natural splendour, with sideshows such as mega-zoos and mega-botanical gardens as remnants of the environment. Deeper environmental ethics are concerned not just with the longer-term welfare of humans, but for the welfare of all else in the environment.

Deep ethics will go further, guiding humans and human egos and selves off centre stage, in cruder versions booting them out of some environments altogether (removing the human *ego* from *eco*). Hypothetical arguments can help in shaking typical human superiority and natural-world arrogance. For instance, humans are confronted, in this way or that, by superior agents, or just more technologically powerful agents, who are inclined to treat humans the way humans tend to treat other creatures and their environments.

The packaging and marketing, selling and implementing of environmental ethics is not simply a matter of promoting environmental matters of interest and concern among an intellectual elite, upper-middle class or other. In important respects, the environment is everyone's problem and everyone's responsibility. The implementation of environmental ethics is a top-down and bottom-up and inside out issue. It is a matter for both the tall poppies and the grassroots. Achieving individual change thereby is a start, but it is not enough. Institutional change is also required. It is not enough that individuals may want to change practices in their own lives. The community in which they live must meet their needs by offering environmentally sound alternatives. For instance, it is virtually impossible for an individual in Australia who is concerned with ozone depletion and the greenhouse effect to buy a refrigerator that does not contain freon.

The implementation of environmental ethics and the provision of environmental options can usefully be compared with overcoming racism. It is not enough that some individuals in the community desire equity, if the laws of the community keep them apart in schools, in pools, and provide separate drinking fountains. The need for an approach that integrates individual and institutional ethical responsibility and environmental concern can be illustrated by recycling. If enough individuals desire recycling facilities, they can usually persuade the community institutions to provide them. Once community institutions provide facilities for recycling and encourage their use, then enough members of the community must use them, if there is to be any real change in attitudes throughout the entire community. To develop environmental ethics actions are needed that influence beliefs and values at individual and community level, where community level may be a village or a nation or several associated nations.

How do we try to motivate appropriate, ethical, action? Set aside inadmissible, normally unethical, means like force, blackmail, bribery, and so forth. Then there are, in particular, two crucial levers: environmental awareness and environmental concern. Environmental awareness is based on information about the environment and influences beliefs about it. Environmental concern is based on emotion and sentiment and influences values concerning it. These elements are interrelated. Beliefs affect values and values affect beliefs. Concern is founded on awareness and awareness is a prerequisite of concern. So there is a need for concern and awareness about both beliefs and values. Belief and values jointly are fundamental to decision making (they appear, perhaps in degenerate forms, with values reduced to preferences for instance, in all models of decision making). And decision making is a prelude to action, and an integral part of rational action.

To alter action, then, by rational methods (as opposed to force, coercion, and so on), we seek to influence beliefs and values, a process which in turn proceeds through information channels, through generating awareness and concern.

The short, but hard, answer to the ways and means of marketing and promoting environmental ethics should also be evident: first of all, concentrate on deeper ethics, and then use every reasonable and ethical way and means available to obtain inculcation and implementation. Interpose into the family, the community, the nation, and the international community as many communication and transaction linkages as feasible, short of overloading, in a beneficial way. Encourage children to be aware and concerned for the environment with a fervour like that for breakfast cereals or bedtime toys and then to operate on their parents with the same enthusiasm that normally obtains the cereal and toys for them. Encourage parents to promote environmental awareness and concern in their children. So that the box of cereal becomes an ethical issue about the environmental impact of what people eat and not just an end product of advertising. Mimic marketing promotional practices, but apply them to the environment, to generate awareness and concern.

No doubt some of the problems with market promotion spill over to ethical promotion. No doubt ethics, being intangible ethereal goods, do not leave a trail of material waste. But there can be problems with their production and about their consumption; in particular, they may not be good products – as narrow traditional ethics are not – and they may have a negative effect on those who consume them – as puritan and work ethics for instance do. We hardly want to encourage the production and consumption of such products. With spiritual goods, as for other goods, efforts to temper unrestrained consumer sovereignty, to increase consumer moderation and consumer awareness, matter.

INCREASING CONSUMER AWARENESS: UNTHINKABLE ZEN AND UNACCOMPLISHED GOVERNMENTAL METHODS

The business of increasing awareness, and generating concern, should be a process in which theory and practice play a game of metaphysical leap-frog. Practice and theory play a sporadic and episodic game of outlook- and attitude-metamorphosing leap-frog. The sporadic and episodic revision of an aspect here and a notion there will produce a cycle of change in outlook and attitudes precipitated by a tactical change following and proceeding and bolstered by strategic change. An action, such as the Gordon-below-Franklin dam protests in Tasmania, questions underlying assumptions about what is and is not possible and appropriate to maintaining environmental quality and the importance of the environment. Aided by a new awareness and new principles for supporting the importance of the environment and its quality, new actions arise, such as further World Heritage listings. The Animal Liberation movement earlier gained notice as an example of the development of environmental ethics arising both out of practice and theory.

> An area in which this game of leap frog can be witnessed is the economics of factory farming. Worldwide, a number of organizations such as Animal Liberation have brought the plight of factory-farmed pigs and battery-caged hens to the attention of the public in an attempt to change both the immediate conditions in which these creatures are treated as biological production-units and the economic circumstances that lead to and promote such production methods. Further fundamental changes to the economic systems of countries like Australia will be needed to establish a new economic order that places animal welfare above profit, habitat protection ahead of wood chips and "thinking like a mountain" instead of thinking like a cash register.[1]

Part of the task of implementing environmental ethics consists in imagining and aiming for what lies entirely beyond the bounds of present practice, thinking the unthinkable. As already explained, ideas such as expanding ethical circles to encompass other species are not new, but for the community at large to take them seriously, to form groups like Animal Liberation, to influence commercial activities such as egg production, is! That is the positive side. There is also a negative side of thinking the previously unthinkable as well. It was thought that humanity could not possibly continue deliberately to destroy its own habitat, that it could not possibly poison the environment to the extent that problems such as acid rain, holes in the ozone layer, and the greenhouse effect, become household words and concerns, let alone that, once problems such as these became known, that humans could continue to produce these effects. True, some steps have been taken, some deadlines and moratoriums declared, but it should be unthinkable that once such problems are recognized that humanity would continue to pursue other equally disastrous courses of action such as felling rainforests, polluting oceans, and using agricultural methods that, for example, in Australia cost 13 kilograms of top soil for every kilogram of grain produced. Who, in their right mind, would eat a hamburger knowing that its production cost half a ton of rainforest biomass?

In times of crisis, throughout history, though cultures and races have vanished, humanity has always more or less muddled through somehow. Whenever a crisis arose or detrimental actions were set into motion a war, a plague or whatever luck or human ingenuity and a sense of self-preservation would stop it, eventually, somehow. But now luck may be running out; now humanity is facing the possibility of irreversibility; muddling through is no longer an adequate procedure. In cases such as the extinction of other species it is easy to comprehend that extinction is forever. Once a species is lost, it does not, it cannot come back. The dodo will never waddle the earth again, nor other hominids such as *Australopithecus*. It is easy to comprehend that once a kilogram of ore is mined from the earth, it will not magically reappear tomorrow to be mined again. Some resources are non-renewable. It is harder to think that deserts may replace rainforests – but 'green deserts' do. Or that even if the rainforests regenerate, at the current rate of their destruction, the world's oxygen supply will be drastically affected before they do.

Part of the unthinkable is that there is not enough time to repair the damage. The regenerative powers and systems of nature may have been so affected that the damage will not be repaired. Humanity could continue to debilitate the environment even in the face of the evidence. The size of human population is increasing so rapidly, and to such a vast extent, that even minimal needs cannot be met without severe environ-

183

mental impacts that would undermine a healthy environment, not to say an already embattled and failing one. There may not be much time; it is doubtful there is enough technology (cf. even the clean-up of large oil spills). Many ecologically well-informed thinkers estimate that there now remain only about ten years to repair the damage and to change environmentally destructive attitudes.[2] After this time it may be too late, trend may become destiny.

The unthinkable must be thought. Forests are more valuable in the longer run standing than as woodchips. Turning trees into discardable cereal boxes, concrete framework, and telephone books is madness. Biological diversity is more important than the relentless preservation of any single species to the detriment of all other species including humans. Humanity is capable of undermining its own support base, and maybe has. Humanity is capable of the destruction of its own habitat, unless it develops an environmental awareness and an environmental concern deeply rooted in an environmental ethic respectful of the environment. Changing to respectful approaches to the environment and supplanting the place of humans in the world and their ethical systems may seem excessive and extreme. Yet what is now seen as unthinkable, as the voice of extremism, will in a decade or two be seen as necessity; what was extreme 10 years ago is now a balanced view.

Deep environmental ethics are part of what was unthinkable two decades ago. To inculcate environmental ethics then, make environmental protection an issue immediate and of vital concern to the populace! There are many now familiar methods for doing this, at various levels of social organization. One method is to enact and publicize legal ways of accomplishing protection; for instance, draft and effect regulations that remove factory farming, that preserve environmental quality, biotic diversity, and a plurality of respectful uses, and such like. Carefully applied methods need not be onerous. In connection with promoting environmental quality for instance, in the USA the Minnesota Mining and Manufacturing Corporation saved $US192 million in less than ten years and "eliminated 10,000 tonnes of water pollutants, 90,000 tonnes of air pollutants, and 140,000 tonnes of sludge after regulatory or operational pressures forced management to focus on waste reduction opportunities".[3] Obvious methods for encouraging environmental quality that follow this same regulatory line of thought include providing tax concessions for non-polluters and heavy financial burdens or closure for unrepentant polluters.

A related top-down governmental method uses standard macroeconomic control mechanisms, monetary and fiscal policy, and so forth. Currently, many countries make environmental devastation financially rewarding. Brazil is given as an example of this by *The Economist* in its summary of a report by Hans Binswanger, an economist with the World Bank. Binswanger reports that in Brazil, its "laws and tax system have made deforestation and ranching in the Amazon artificially profitable".[4] Brazil has encouraged deforestation by prodigal fiscal and monetary policies leading to high inflation; by exemption of agriculture from taxes; by land taxes on unimproved land, i.e., rainforests. Furthermore, government agencies give tax credits for investments in approved schemes in the Amazon; government subsidizes rural credit. In addition, laws on squatters' rights are perverse disincentives for conservation. To top it off, there are few forest guards and those few are vulnerable to bribery by business agents and other smart operators. Correcting this lamentable situation calls for some simple, but hard policies.

> Brazil should tighten its fiscal and monetary policies, so that investments in financial securities become more attractive and land investment less. Tax exemption for agriculture needs to be phased out so that farmland is no longer used as a tax shelter. Land taxes should be reduced in the forest in order to encourage conservation. Tax credits for cultivation and ranching in the Amazon should be revoked. Rural-credit subsidies should also go, for ranching if not for all agriculture. Squatters' rights should be limited to perhaps 100 hectares rather than 3,000. A ceiling should be put on corporate holdings in the forests. And forest guards should be given a share of all fines levied on those breaking forest rules in order to provide an incentive to catch trespassers rather than to accept bribes.[5]

An almost reverse suggestion has been made by Passmore, a suggestion that would certainly aggrevate the situation in Brazil. With regard to pollution at least, he suggests that instead of penalizing the polluters, governments should subsidize the polluters' efforts to rectify the situation and to clean up existing damage. Passmore makes several points about anti-pollution measures, which include "that ecological problems are social problems, not scientific problems; to solve them satisfactorily is, in most instances, to be faced by a sub-set of problems, scientific, technological, economic, moral, political, administrative" and "that any proposed solution, to be satisfactory, to be 'operational', must take into account, on a wider scale than has normally been attempted, the costs and the benefits resulting from the use of that, or another, method of control".[6] By costs and benefits, he means more than financial costs and benefits. In particular he means social costs and benefits. Passmore holds, on the one hand, that it is wrong for anyone to poison the environment of another, and, on the other hand, that anyone who does poison the environment of another is also poisoning his or her own environment:

> The thought of subsidizing the polluter, paying him to make use of substitute materials or to install anti-pollution devices, horrifies the more extreme sort of retributionist. The polluter, as the retributionist sees the situation, has wickedly poisoned the air and the sea; let him now pay for his misdeeds.[7]

A point that Passmore is trying to argue is that 'we are all in this together', or better, that social problems generally require social solutions. However what Passmore apparently fails to see, is firstly that – whether one is a retributionist or not, whether one confuses blame with retribution or not – by subsidizing the polluter the rest of us have to pay twice for cleaning up the mess; and, secondly (along with development orthodoxy), that the polluter is not part of the solution, only an enhancement of the problem. Naturally this line of criticism is held by many to be dead wrong with the emphasis on 'dead'. What environmental philosophy can contribute is an understanding of obligations not to poison the environment and obligations to rectify the problems. From both an economic and governmental point of view, intelligent use should be made of appropriate mixes of regulatory and legal mechanisms and economic mechanisms to obtain desired results. With a new and different legal and regulatory framework and proper resources accounting, environmentally desired results could even sometimes be achieved through market forces, redirected to promote preservation of environmental quality. Part of what is required is clever structured adjustment and provision of incentives. In a similar way there needs to be adjustment and resetting of targets, so as to make environmental considerations an integrated part of economic and public (or governmental) planning. But a prime adjustment has to be in ethics and ethical practice.

Another different method, but again partly governmental, proceeds through education. For example, it aims to promote environmental education, so that, above all, increased awareness and concern result. Such environmental education can no longer satisfactorily avoid, at less elementary levels, environmental ethics and issues therein. In Australia, there are occasional courses in environmental ethics at some universities and colleges of advanced education. But most do not have them, and some that did no longer do. At the university level, however, the number reached is small and many of those who take such courses are already converted to environmentally responsible attitudes. Environmental education including ethics, needs to be much more extensive; it needs to begin at the primary level and should not be confined to formal educational structures. First then, environmental ethics should be included in 'philosophy for children' curricula, where they occur (which should be more widely). If, moreover, as the South Australian Department of Environment and Planning claims, "Environmental education seeks to develop students' attitudes and values by fostering their understanding of environmental issues", then advantage should be taken of media coverage of environmental issues not only to display the issues, but to foster an understanding of environmental ethical and philosophical issues underlying them.[8] This would require intensive community level education programs and peer group support to encourage people to adopt a positive environmental ethic, greener ethics. What is required is to change fundamental attitudes so that everyday matters, such as what a person eats and excretes, become ethical and environmental issues. People must understand that if they live as if the environment matters and it does not matter, then it does not matter, but if they live as if it does not matter and it does, then it does matter – it matters a great deal. Furthermore, they must understand that it *does* matter. Environmental concern needs to be made a basic part of curriculum at every level, plus making use of mass media to supply and promote environmental education. More aptly, education needs to be appended to environmental awareness rather than environmental awareness added as another course to the curriculum, just as environmental ethics is more than another section of applied ethics. Environmental education is more than merely plugging new subject matter into teaching programs. Environmental education includes teaching and experience of lifestyle survival skills and much more. The matter bears elaboration.

SETTING COURSE FOR DEEP ENVIRONMENTAL EDUCATION

Education is overwhelmingly important for an extensive range of environmental issues. It is perhaps enough to consider two main components of environmental impact equations, population and consumption; the bearing of education on the improvement of technology and management of environmentally superior technology is evident. As to population, increases in the level of education of women are significantly correlates with decline in human birth rates. Indeed education of women may be the best single indicator of percentage population growth rates presently available. Improving education, its extent and quality, along with making satisfactory birth control technologies widely and cheaply available, appears to be the major method for emancipated human population limitation. Of course other things matter as well, other factors enter, such as improved social security, but these factors are apparently less decisive for growth rate decline and certainly more difficult to deliver. By contrast, industrial growth, though much acclaimed as *the* route to demographic transition, only features because it tends to carry and aggregate those features that do matter, and because it may enable education , social security and crucial services, to be offered (though with usual countervailing externalities). Further, industrial development of the regularly tendered unselective sort clashes with reduced and careful consumption, the remaining major impact component where education can again make a major difference. Education can be enlightening: it can supply a wealth of information concerning goods and services and their comparative impacts. Education can be enabling: it enables further information to be accessed about environmental features of goods and services, about how to tread more lightly on the Earth, and so on.

Evidently education also bears on many other environmental issues, less directly linked to impact components; for instance, such issues as habitat retention, biodiversity, and so forth. As one example, education has a central role to play in trying to alter exploitative attitudes to habitat or wildlife in many African villages. In the present situation, apparently easy money from an Asian-sponsored ivory trade has induced many villagers, succumbing to monetary temptations, to annihilate animals. Present short-term policy, in parts of Zimbabwe, for instance, consists in buying off the present generation of exploiters, with alternative money; payments for conservation, together perhaps with penalties for failure. A sounder affordable longer term option includes educating future generations in conservation practices. In education lies a main hope for the future.

While it may look difficult to overstate the importance of education, education is, however, far from a panacea, even in the longer term. That education is no panacea, no unqualified environmental saviour, can be gauged from respective environmental impacts of various formally educated and formally uneducated groups of people: Anglo-Australian as contrasted with Australian Aboriginals, American Jews as compared with American Amish. Rather *right* education, whether formal or not, is part of a larger adjustment package. Right education is enlightening and enabling. More of the wrong sort of education (which includes much of what is presently dished out) will tend to compound environmental and other problems. Education may be directed at dubious or defective objectives; and it may not succeed in producing or promoting intended objectives, for many reasons. For example, as with mere rote learning or detached theory, it is not absorbed, adapted or applied. Worse, what is commonplace, what overlaps wrong direction, it may amount primarily to indoctrination, indoctrination into a particular culture, typically into a prevailing social paradigm. Education is typically ideology bound and ideology reinforcing. It is an essential element in maintaining the dominant industrial paradigm. Almost needless to add, these frameworks may not be environmentally friendly at more than a superficial level. The dominant Western paradigm into which most of these humans with regular access to consumer goods and television are now inducted remains environmentally hostile, beneath a patchy veneer of acceptable change.

Education is thus two-faced. It frequently faces back to established ways and ideologies, implicated in present environmental crises, or even further back to, perhaps environmentally more benign, religious darkness. Education in a fundamental religion or a power ideology affords a familiar example of the latter sort. While a few so indoctrinated later escape, often as intellectually handicapped, most do not. With the former similarly, education leaves its heavy mark. For an instance of the former, consider new business sponsorship of education in the USA (displacing local government funding). Its avowed objective is to induct students into capitalism, its success and its practice, and into associated consumerism and business values – much of its still older anti-environmental, anti-social business, too much of it still ugly American or MBA style enterprise.

Even where education faces forward, as it is presumed to do under contemporary educational liberalism, it may remain essentially anthropocentric and unduly shallow, as happens under technological, emancipatory and even neo-romantic educational agendas.[9] Even much of what parades under the banner of 'environmental education' is shallow, some does not even rank as environmental at all. Such frequent themes as "providing for equity within and between human generations" are shallow, primarily social, and only oblique to environmental issues. As with so-called 'environmental economics', much would be more accurately presented under a *resource* heading. What is being offered is, too often, resource education.

While there are many types of education, *most* types are not environmentally satisfactory. Education in itself is no remedy for environmental vandalism, any more than for racism, hooliganism, or animal abuse. It has again to be a right education, coupled with right practice – or more colourfully "pigged-headed in the right direction".

A right sort of education is, in turn, now an environmental one, most preferably a deep one. A right education does not reinforce old superstitions or prejudices; but nor will it underwrite contemporary dominant environmentally hostile paradigms and practices.

Unremarkably, it is much easier to point out what makes for an unsatisfactory education environmentally – much of what is presently on offer around the globe will serve – than it is to indicate what would belong in a right education.[10] But we can indicate a right education straightaway, then backfill on details. A right education is a *duly pluralistic education, in conformity with a satisfactory culture* or ideology, in conformity with an environmentally adequate paradigm.

Educations can be helpfully classified according to their governing paradigm, ideology or culture. Thus, for example, Catholic, business, and Aboriginal educations, themselves of various kinds depending on the brand of Catholicism, the style of business, the Aboriginal grouping. There is, of course, much overlap. As a Christian Brothers education is a sort of Catholic education, so, by and large, are Catholic educations part of Western educations conforming to the dominant social paradigm. Structural relations of educations *in* paradigms derive from structures of the paradigms concerned. Classifications likewise transpose; for example shallow educations are educations predominantly within shallow paradigms.

While most traditional educations were educations in a culture, in a local culture, and most contemporary educations remain predominantly that sort of way in a larger culture – educations *in* a received ideas-systems, answering to state, business or religious orders – not all education, even tribal, operates that way, none needs to operate that way, and all could be improved by not operating just in that way. All could be improved by sympathetic introductions – or more – to genuine alternatives, something that sometimes happens in language, anthropology and social science courses. But it tends not to happen in central courses of curricula where core cultural values are usually reinforced, and alternatives not offered, or even glimpsed by students. That should change: education processes should engage with alternatives. Teachers should teach about alternatives, students should learn alternatives.

There are then two types of major changes we are suggesting in education: pluralistic as well as environmental. Part of the point of such pluralism is evident enough. A pluralistic education can encourage tolerance, and reduce bigotry, racism, speciesism, chauvinism, and so on, can enable a more sympathetic approach to other cultures and other

191

creatures, and can give recessive cultures a chance. Another part serves environmental ends. Even if but a shallow environmental education is achieved in many more places – still, with pluralistic environmental educational arrangements many more could become acquainted with deeper alternatives.

Because educations in recessive or minor paradigms almost invariably involve or lead to working acquaintance with other, especially with dominant paradigms, we can, for most practical purposes, omit the qualification to pluralistic education in explaining deeper education. (It can be conceded that in different unlikely circumstances, a deeper education that was not pluralistic could also be narrow and doctrinaire, and could perhaps even foster socially undesirable outcomes.)

Featuring in what is served up in dominant shallow education is the following rickety framework: modern liberal democratic political arrangements (namely, market capitalism with private property and representative government) are just fine, indeed close to the best possible, and require at most minor adjustment (e.g. fine tuning). With the demise of state socialism, the practical quest for alternative ideologies has been properly exhausted and has effectively ended! Thus too there is now an end to ideology. There is accordingly no need to consider alternatives or to change basic values, those of possessive individualism and therewith capitalism and consumerism, or basic structures, for instance class and power hierarchies, and prevailing human chauvinism. Economic growth can continue more or less as usual, and must (to alleviate persistent national unemployment, then perhaps third world poverty and other shallow causes). Business can proceed much as usual, with minor changes (most of which make good business sense) to remove worst environmental offences. Most of these changes can be accomplished by techno-fix, technological fixes to modify or moderate industrial impacts, coupled with improved management. Moreover, sound management and regulation will ensure that technology is properly applied; stewardship, its analogue for land, the earth and natural resources generally, will ensure similar control for the earth. Nature itself, which is merely a backdrop to human affairs on centre stage, serves primarily as a source and storehouse of raw materials, which have worth only when humans value them. Their value is established or enhanced when humans extract or shape, mine or make things from them, for then with human labour mixed in they become commodities with dollar value. Perhaps natural items also have some value when humans worry about their continued existence and would be prepared to pay good dollars for their retention.

In brief, what the dominant shallow paradigm asserts is what shallow education teaches. Instruction mirrors ideology. In most of the West that ideology includes capitalism, main (more apparently benign) themes of which pervade educational curricula: healthy growth, accumulation, consumption, power, strength, control, and so on (highly problematic and deleterious themes one and all, whose negative impacts have to be got across decisively in rectified educational processes).[11] The basic purpose of such education, apart from further attuning people to broad features of the shallow paradigm, is to train them to fit into prevailing production and consumption patterns, to fit them out for a disciplined work force, with the industrially relevant expertise there demanded, and to turn them into dedicated consumers.

While features of shallow education have been assembled and tellingly criticised in accessible literature,[12] regrettably the same sort of attention to detail cannot be recorded for deeper education. Asking about what a deeper education would look like tends to draw responses that are generously described as mush and waffle, though redolent with wholeness and naturalness: stuff and (often) nonsense, which even if sometimes suggestive, teachers could not teach to ordinary students. Such stuff includes, what might be admirable enough in certain settings: developing the whole person, affording opportunities to attain a total view, developing capacity to achieve self-realization, rejecting compartmentalization and excessive analysis, diminishing divisions (of labour, etc.) and hierarchies (of power, etc.), encouraging deep questioning, promoting comprehensive love and understanding (e.g., as a substitute for compulsive or excessive consumerism), instilling a spiritual dimension, and so on.[13]

Yet it is moderately clear what would be taught in a deeper education. It would be both pluralistic and environmental, environmental in a deep fashion, and so highly critical of present arrangements, by contrast with prevailing education. As regards pluralism, it would offer genuine and serious alternatives, all the way through a curriculum, from ideal sciences outwards. Where geometry is taught, for instance, teachers would endeavour to give some impression at least of alternatives, and what living under them (as perhaps we do) might be like; similarly for arithmetic; and so on.[14] Of course, it would investigate the shallow alternative most of us labour under and critically examine themes and assumptions of the governing social paradigm and its important variants.

As regards environmental education, elements of deep alternatives would naturally be taught, instruction would again follow ideology. But these elements would not figure only in ecology and geography streams (which themselves would be significantly different: less analytic, reductionistic, managerial in approach, less celebratory of human achievements and economic growth, etc.) There would be major changes in economics and accounting and management, in social and political studies. Nor would backwatered humanities and cultural studies, nor arts and crafts pass largely unchanged. Gatherings on literature, film, television and other communication and artistic media would not merely be apprised of, but closely acquainted with, the shallowness and chauvinism, speciesism and humanistic prejudice, of most past works and offerings, and would gain some appreciation of what different deeper media, uncluttered by human stars, would be like. (Recent, deeper and more realistic, approaches to past glorious wars, and to display and acclamation of military virtues, afford significant working examples.)

A crucial component of this new education would be far-ranging biohistories,[15] green histories of the earth and other planets, disclosing in particular what humans have done to the earth, its systems and habitats, the often astonishing ideologies they operated under, and what alternatives they had that were not taken. Coupled with green history would be a much more speculative subject, green futurology, exposing possible future paths and relevant action in averting deleterious courses.

A deep education would also be a critical education, not merely of shallowness, but self-critical as well. For example, looking out, it could examine and dispatch many prevailing myths (many deriving from dominant paradigms), both of general character and specific to distinct environmental topics, such as forestry. As these myths are explained and dispelled in accessible sources, perhaps it is enough to sample a few types here. There is, for instance, the general myth that there are no unsolvable problems, in particular no insoluble environmental problems. Science and technology will always find a way (the myth of universal techno-fix, so to say). Closely linked is the myth that everything is accessible to knowledge, nothing is in principle unknowable; ignorance is entirely removable. It too is a solvable problem. On this myth was based the Cartesian doctrine of a universal method. Given the Baconian linkage of knowledge with power through control, the first bundle of myths leads to the idea of total control. This transports us to another bundle of myths, that with enough knowledge and technology humans can completely 'manage planet Earth'.[16]

By contrast with the curricula, methods, means and many materials for environmental education are by now moderately well appreciated. Means include communicational means of all ethical sorts, not merely books, newspapers and television, but theatre and creative arts, exhibitions, debates, processions, wakes and celebrations. Methods, which overlap means, include case studies, projects, problem solving, structural analyses, experiments, learning by doing, copying, simulations and modellings, field trips, and so on. Materials, overlapping again, include blackboards, notebooks, guides, textbooks, instruction manuals, posters, software, and so on.

But naturally education should not be confined to classrooms, educational institutions or stereotyped settings. It should occur in daily practice, on the factory floor and in field and forest. Nor should it be equated with or terminate with formal education. There needs to be continuing, continually adapting, education.

Deeper education too has its limits. Despite the advantages of extensive and intensive educational programs, environmental ethics cannot really be got across just by teaching, still less by preaching. If they were, it is unlikely that they would be accepted, as the example of the resistance of foresters in Tasmania shows. Simply presenting them under that label may encounter reluctance, or reaction, and thus be counterproductive. Presentation of information is as important as the information presented. Values should be inculcated indirectly. Environmental ethics are more likely to be effective if presented indirectly, through immersion field trips to impressive areas, positive *engagement* in relevant practice, and so on. Good television and radio programmes with the right messages worked in can be important sources, but it is better to remove ignorance by getting people out and involved than by simply trying to convert them with images and rhetoric. The obvious must be made inescapable. Beach users in Sydney staged a massive clean-up campaign when (in January 1989) they could no longer accept the level of litter on their beaches. They removed 3000 tonnes of litter in one day. What was obvious was that beaches were no longer pleasant – or healthy or even safely useable – because of the refuse millions of Sydneysiders had thrown away. What was not obvious was that there is *no away*. Part of the role of environmental ethics is to show that there is no away, that the 3000 tonnes of litter collected from the beaches have only been moved to some place else. Also it is part of the role of environmental ethics to show that there are ethical issues involved with the production and ultimate resting place of the items that made up those 3000 tonnes of litter.

MODES OF ENVIRONMENTAL MESSAGES, SUCH AS HUMOUR AND NOVELTY

As well as method, the mode and format of the method and message matter. Serious and important messages can be communicated in less than serious formats. An indirect means of presenting environmental ethics is through humour. Humour can bring ideas to a person's attention that might otherwise be overlooked, ignored or shunned. Humour can get under a people's intellectual guard and confront them with ideas that they might not otherwise consider or accept. Humour does not always hit its mark, but it does so often enough to be a good device for conveying new ideas or challenging old ones. Humour is a useful tool for negotiating a common ground between the speaker and the audience. Witty remarks advance social discourse, foster group cohesiveness and strengthen social bonds. In other words, humour can bring another person around to the speaker's point of view and do it painlessly. When the topic of the humour is the environment and it is presenting a case for the preservation of the environment, the speaker can cajole the audience into group cohesiveness.

A prime example of using a humorous, superficially non-serious messages occurs in the writings of Douglas Adams. His books such as

Hitch Hiker's Guide to the Galaxy and *Restaurant at the End of the Universe*
are filled with serious messages conveyed with humour. Adams uses both
style and content to put across a number of environmental messages.
While many of his passages concerning the environment are too long to
give here as examples, the following passage gives some idea of how
substantial messages of environmental ethics can be conveyed through
vulgar humour.

> The fabulously beautiful planet Bethselamin is now so worried about the
> cumulative erosion by ten billion visiting tourists a year that any net
> imbalance between the amount you eat and the amount you excrete
> whilst on the planet is surgically removed from your body weight when
> you leave: so every time you go to the lavatory there it is vitally important
> to get a receipt.[17]

In countries like Australia and the United States that have huge erosion
problems, the message will strike home to some. If the humour indicates
an environmental ethic that emphasizes consideration of the needs of the
ecosphere along with those of humans, the connection of everything to
everything else and the need to remain in harmony and balance with
nature, then a direction in which to look for a possible solution would
be part of a palatable presentation.

Of course humour does not always successfully fulfil these intended (or one unintended) functions. Sometimes it misses its mark.
Those without senses of humour may fail to see what is funny about the
city council putting a by-pass through their home. Those with a zealot's
turn of mind may not fully appreciate the loss of a million species by the
year 2000. A farmer may not get the joke that in Australia for every loaf
of bread produced seven kilograms of top soil are lost. Also because
humour is culture specific, it is necessary to select forms of humour that
will appeal to the target audience. Cartoons will work in some cultures
and with some people, but totally miss the mark with others, with whom
however humorous stories will succeed. But whatever the form, humour
is often an excellent vehicle for helping to deliver elements of an
environmental ethic.

Allied to the use of humour is the use of novelty. Ethics without
tears can be delivered by presenting the message in a novel manner. Take,

for instance, the idea of chocolate chip mining used in *Project Outlook* in New South Wales. Students are issued a chocolate chip biscuit (cookie) and instructed to remove the chocolate chips from it. They measure the amount of chocolate chips 'mined' and are then told to reconstruct the remains of the biscuit back into a whole again. The students are asked, "How is biscuit mining like real mining?" From this exercise the student learns that minerals are not evenly distributed and that mining dramatically affects the Earth's surface.

What is so clever about this technique is that if the students understand that there is a relationship between what they have done to this familiar object and what is being done to the Earth's surface, then they will also realize the difficulty of reversing what is being done. The devastation caused by mining will be real to them. Here – before them – on a scale that they can comprehend is the same process and effect that is occurring on a larger scale at mining sites. It is not difficult to pack into this lesson elements of an environmental ethic that makes decisions about mining more than a trivial affair.

RENDERING ETHICAL WAYS AND MEANS INTEGRAL TO PRACTICE

A further method of promoting environmental ethics is making them important to those at the places where environmental crunches come. For instance businesses and industries that pollute or despoil the environment can be shown ways of making environmental ethics profitable. Companies such as 3M have found that it is profitable to promote environmental issues. 3M has started programs such as "Pollution Prevention Pays". The promotion of environmental quality and developing a positive environmental ethic is frequently not counterproductive to companies, but makes good business sense, both for the money saved and for the publicity value. Where it is not, that enterprise should be appropriately adjusted or else eventually wound down. The same is true for agriculturalists, who have for years polluted the soil, water, and sometimes the very products of their business. Among others Rachael Carson warned for years of silent spring and of the effects of pesticides.

It cannot be in the long term interests of agriculturalists to destroy their own livelihood.

Those interests which are damaging the environment and threatening the quality of life, livelihoods, and indeed the very lives of humans and other species, also threaten themselves by their own actions. It is important to replace the 'pioneering ethic' – with its mentality of develop at any cost – with an ethic more compatible with continued and sustainable environmental quality. A way of doing this is to impress upon the parties concerned that other avenues are open, or can be opened, and that at least some of these avenues are environmentally sound. It is therefore in their own longer term 'enlightened self-interest' to develop integrated practices which maintain their own livelihoods and at the same time encourage and protect a healthy environment. Take for instance the marine fishing industry in Australia. Two main recommendations from the Australian Fisheries Council highlight the need for parties directly involved with the use of what should be renewable resources to reconsider their environmental impact and improve their communicational activities. The recommendations are:

- That all parties associated with the marine environment form working groups to ensure an orderly, balanced approach to development of the marine environment.

- That increased effort be put into education, research into the worth of the marine environment, into the practicality of implementing environmental restoration and regeneration processes and the development of a database to heighten public knowledge.[18]

Of course compromise ('balance') and communication are no substitute for rectification of the underlying environmental problems, greedy overfishing of 'the commons'. But once such groups as the Australian fishing industry have recognized the need for environmental awareness and concern, their acclaimed desires for orderly and balanced approaches and their increased efforts in education, do offer opportunities for introducing and pushing environmental ethics, for instance, encouraging a properly ethical approach to overdevelopment problems. Primary

199

industries, such as the fishing industry, timber industry, and others dependent on 'renewable resources' have literally 'primed' themselves for environmental re-education. In fields such as medicine and veterinary science, ethicists (duly ideologically culled) have increasingly become part of decision-making and education processes. The same could apply to areas and organizations involved in what should be sustainable primary industries.

Along related lines, elements of environmental ethics could be built into environmental impact statements (EIS). For instance, EIS processes and associated cost-benefit and other analyses, should be made genuinely ethical, not merely aligned to conform with legal requirements (which of course diverge from the ethical). More, they could be deepened, to take some account of intrinsic values, and not merely questionable hypothetical dollar pricing, and they could be ethically constrained, e.g. to avoid infringement of creature rights. But such deeper economic assessments of development projects, environmental ethical assessments, for all that they are feasible, still appear some way down political tracks. Even so more could be achieved of ethical relevance. In many countries, EISs are required for major projects, but are considered by developers as little more than perfunctory formalities, carrying little of the intended weight or the effects upon awareness and concern that brought them into being. If EISs were a precondition for loans, and if they were not only required, but required to be produced by properly independent outside bodies in a manner like external audits, at the expense of the party wishing to build, develop or otherwise impact the environment, then an environmental ethic could be implanted both in the requirement to have an EIS and within the parameters of the EIS. An EIS would constitute more than a bureaucratic nuisance, as they are seen now, and would be a necessary part of the argument for permission to build, to develop or otherwise to disrupt the environment. Just as a project does not go ahead until finance has been arranged, so could it be with environmental clauses that were taken more seriously than minor impediments to overcome, mere environmental windowdressing.

Changing use profiles is another method of incorporating and promoting environmental ethics. Two examples of possible room for change can be drawn from the city of Sydney. On current estimates each person in Sydney uses 5 litres of water to brush his or her teeth. For the conservation of water, a precious resource in Australia, toothpaste and tooth brush companies and advertising agencies could be offered incentives, tax exemptions or reductions, for instance, for showing ways of brushing teeth that conserve water. Other possibilities exist for changing water use profiles, with user-pays practices much favoured under the prevailing economic paradigms. A few examples of other ways of changing water use profiles include separating 'used' water from different domestic uses, recycling of grey water and so on; source point anti-pollution measures; new uses for 'used' water, such as fertilization; changing irrigation practices; converting to ecological (dry) toilets instead of flush toilets; and generally stopping use of water where it is not needed, such as car washing and lawn watering. Local councils in Sydney are presently promoting the use of big bins for garbage. To conserve resources in the form of recyclable items – glass, paper, metal, etc. – the use of indiscriminate source point refuse collection, such as big bins should be discouraged. Point of origin recycling is to be encouraged, instead of dumping of potential recyclable resources into one amorphous mass. The ecological benefits of recycling include good resource management, reduced energy consumption, and reduced pollution. Consumption and disposal patterns are *ethical* concerns; for the ways humans obtain resources, how they conserve resources, and how they dispose of former resources, have implications for the way other species, other humans, and the environment is treated. Flagrant disregard for these concerns is tantamount to ethical negligence. An environmental ethical point to be made here is that refuse (and in some cases pollution) are, among other things, resources. Reusable and recyclable items are substantial sources of perhaps much needed materials. It is a change of attitude within the 'disposable cultures' that is required, and environmental ethics can indicate needed changes and the direction they should take.

A way of stating the ethical concern involving negligent use profiles draws on the fashionable environmental slogan, 'Think globally, act locally'. Much of the environmental destruction in Third World countries and developing economies results from pressures placed on those countries and economies by developed countries and economies.

The American fast foods market is helping to convert Central and South American rainforests into hamburgers. If Americans (and Australians) had to shoulder the costs for the environmental destruction caused by their trivial pursuits, then they might take better care of the environment beyond their own backyard. They might also take responsibility for it.

A local way of taking better care of the environment and ensuring some responsibility for it would also hopefully involve channelling benefits to the people and areas that do the work and make the sacrifices. Consider for example the Amazonian forest guards (mentioned by Binswanger). If they had a stake in the preservation of the forest they would be more likely to foster its preservation, particularly if there was more profit directed to them in keeping it standing. Another example can be drawn from Kenya. Tourists pay large sums, mostly in Nairobi, to venture to the various reserves, where they pay additional sums, mostly for drinks and souvenirs, to the management of the lodges and camps owned by the interests they paid in Nairobi. Most of the tourists' money remains in or winds up in Nairobi. While some money does make it to the provincial towns, still less goes to the people, who need it, on the edge – on the edge of reserves, on the edge of poverty, and on the edge of subsistence. It is the Kenyans who are on the verge of despoiling the parks and poaching the wildlife who need to receive a greater portion of the financial rewards of tourism. Some Masai are hired to guard camps; some sell trinkets to the tourists; but few make any real profit or gain any meaningful share of the tourist expenditure even in Kenyan terms. Where communities so far unmoved by deeper environmental concerns do not participate in the management of species, they may feel no responsibility placed on them for the continuance of the species.

Changing attitudes towards recycling and changing use profiles are two steps in more general methods of promoting environmental ethics – first through revealing how ethics enter (and are bound to enter in some way) into environmental decision-making and action, and second through changing demands on the environment to make quality of life and not diversity of commercial baubles the driving force. Quality of life should not be confused with style of living. Lives that are "Simple in means and rich in ends" provide quality, rather than lives that are economically governed and perhaps glutted with hollow trappings of commercialism. This in turn involves such matters as the promotion of steady state economies over growth economies:

> environmental problems have a common source. And it turns out to be the same source of our social problems. It is not the pursuit of economic growth and an ever-increasing material standard of living. It is the pursuit of these things to the virtual exclusion of all else. We are putting too much emphasis on wealth generation and not enough on its distribution and conservation.
>
> The links between economic growth and environmental degradation are clear. Japan's striking success in becoming one of the world's biggest economies has come at the price of becoming also one of the world's biggest polluters.[19]

Richard Eckersley, a principal issue analyst for the Commonwealth Scientific and Industrial Research Organization, goes on to comment that the need to integrate economic and environmental objectives to achieve sustainable growth, sustainable society and broader, longer-term perspectives "will require far-reaching changes in the way we manage our economies.... The reconciliation and integration of all those goals – economic, social and environmental – is the most important challenge facing this country, and the world".[20] Unfortunately, like so many commentators, he does not get far beyond these contemporary platitudes, beyond vague generalities. But we can.

Some opportunities to practice responsible attitudes towards these and those environments, instantiating and promoting environmental ethics, are in place; others need only be re-oriented towards the environment, and away from exploitative practices; yet others need to be instituted. While almost everyone can do something, much requires dedicated people; and in these individualistic times, leaders and exemplars can play important parts. For example, charismatic persons can promote environmental ethics. Charismatic ecophilosophers, where they can be found, and charismatic persons with a sound grounding in environmental ethics, where they cannot, can influence the attitudes of others. Among key influencers in the latter category are film and entertainment stars (environmental 'actor-vists'), members of royal families, politicians and comedians, television personalities and local individuals. Among the film stars who have made commitments to preserving the environment and persuading others to follow suit are Brigette Bardot, Robert Redford, Sting, Jack Thompson and Peter Ustinov. Members of royal families whose names are associated with environmental issues are Prince Bernhard of The Netherlands and Prince Philip, the Duke of Edinburgh, UK, both of whom were patrons of the World Wildlife Fund and Prince Philip is now the president of the former World Wildlife Fund now expanded to the World Wide Fund for Nature. Charles, the Prince of Wales, is also heavily involved with environmental issues.

Besides individuals, all of clubs and groups, factions and organizations, can realign their attitudes and influence those upon whom they impinge. One important example of this is the beginning of a shift in attitude by the World Bank about the kinds of projects the Bank supports and the environmental soundness of those projects. The World Bank was for years an institution that promoted environmental destruction with its lending policies. Although it has by no means totally repented, it has begun to question the environmental soundness as well as the financial value of some of its investment policies. Both the countries which exploit the environmental resources of other nations and those countries exploited can have pressure brought to bear on them by a better environment policy and revised ethic from the World Bank. Similarly for many other purportedly international institutions (e.g., IMF, GATT, UNESCO, etc.); ethical and environmental quality controls need to be internalized in all their practices.

8

Suggestions on a Range of Initiatives and for Action

Just as inculcation and implementation of environmental ethics can take place at various levels of organization – from individuals, through small groups such as families, clans and clubs, to progressively larger groups such as councils, unions, armies, federations and states, and finally to transnational groupings such as churches companies, banks and United Nations organizations – so initiatives and action can be taken at each of these levels. While each of these groups could, and should, reorient their commitments and practices, both ethically and environmentally, some of the suggestions for adjustment and reorientation that apply to small groups make no sense for larger organizations, and conversely. For instance, lifestyle changes, which are important at individual and family levels, do not significantly apply to larger organizations such as states. Conversely, constitutional legal changes and macroeconomic reorientation which matter for various regional or state initiatives, though they affect individuals, do not significantly transfer to individuals. A rough table, plotting levels of organization by size one way, and presenting areas of possible action (such as lifestyle, education, entertainment, economic, politics, law) the other way, and mapping extent of significant applicability, is instructive. Such a rough undrafted table looms in the background in what follows; it helps to structure the discussion.

At every organizational level two interwoven types of change are called for, and should be sought. First, to begin systematically: to amend everyday practices to render them environmentally sound and friendly. Second: to alter the background ethos; to introduce elements of the informing environmental ethic behind those adjustments and modifications, to guide, justify and strengthen practice. An ethics guides and informs, justifies and strengthens practice in ways that a mere economics for instance cannot. We begin with suggestions for changes applicable only or primarily to independent individuals, and then work up through organizational levels.

205

Chapter 8

INDIVIDUAL AND SMALL GROUP INITIATIVES AND
ADJUSTMENTS

There are a great many changes individuals, especially richer humans, could well make to their lifestyles, which would make a difference. For certain worthwhile results it would take many individuals making a concerted effort, but in some cases it would not take very many individuals, but only a few who served as conspicuous examples or role models. At the individual level many of the requisite types of changes are well-known, and many of the handbooks setting them out have received wide circulation[1]. So we can be brief, simply listing main features with some commentary. The environmental impact formula applies to individuals as well as groups such as families, firms and communities: an individual's impact will turn on his or her consumption as weighted by the technology involved. Individuals who have room for choice, richer individuals in all countries, and a great many individuals in richer regions, should aim to adjust their consumption and, so far as feasible, to improve their technology appropriately. Such lifestyle practice is generally enjoined by environmental ethics, which would have individuals tread more gently upon the Earth. But it runs counter to the main tenor and ethos of contemporary living, which is substantially committed to more conspicuously consumptive and extravagant lifestyles: bigger houses, more powerful automobiles, more consumer goods, further material status symbols, and so on.

Reduction in consumption of some commodities is particularly important: ivory, whale oil, big-cat skins, because of their effect on remaining samples of large animals; CFCs, such as freon, because of their impact on the Earth's atmosphere, the ozone layer especially; non-renewable energy supplies, such as fossil fuels, because of their pollution effects and contribution to greenhouse heating; and so on, through a substantial list. But there are sometimes serious obstacles in the way of such lifestyle adjustments. For the most part refrigeration units without CFCs are not available; an individual who would avoid CFCs entirely faces the difficulty of life without a refrigerator, and thereby given a fairly orthodox city life, minor hardship and additional lifestyle complications. With energy consumption there are related difficulties; a person may have to run an automobile to retain a job or to commute to it, may have

206

to use fossil fuel directly or indirectly to heat a poorly insulated and sited house. Many individuals, even in the richest countries, are locked into consumptive lifestyles, from which they have no easy escape, even should ethical messages convince them they should make some effort. But social arrangements are stacked against them, especially if they are not isolated but carry family commitments (thus the escape barriers, operating against rather small but meaningful environmental lifestyle changes, are surprisingly high). New appropriate lifestyles, *green* lifestyles, should be made feasible, and preferably easy; but there are serious complications in achieving requisite structural change. One regular problem encountered is the *inaction circle*, interlinking individuals and planners. An individual can only effect significant lifestyle changes given changes in social arrangements; but social planners, politicians and others who might help achieve requisite social adjustments, observing individuals revealed preferences and their lack of effort to make such changes, conclude (much too quickly, but very willingly because of their commitments to prevailing arrangements) that no such changes are really desired, individuals 'vote with their feet'; so nothing changes. For example, there is substantial evidence, both direct and indirect, that very many city dwellers in the more affluent West would much prefer to live outside cities, in quieter more peaceful rural settings. (For example, a questionnaire in Swiss cities disclosed as much.) But mostly they have little or no opportunity to do so. So they remain (stuck) in the cities which is where their activities "reveal" they want to be.

Practical action, to be effective, must go a great deal further than joining a local birdwatching or bushwalking organization – though that can be important for consciousness raising, and though several of these types of organizations are now making political efforts of an environmental sort they would not have considered a mere 10-15 years ago. Undertaking (or, still less adequate, contemplating) birdwatching, bushwalking or basketweaving may be enough for (utterly) lax green action, but for genuine greenness more is expected of individuals, considerably more, both on their own and in their normal social settings. Much practical action by individuals, to serve environmental imperatives, is inseparable from householder activity, most commonly family action. A prime objective of such local action is environmentally sound housekeeping, in the generalized sense, sound ecological living. It includes such things as:

- reducing or eliminating consumption of items which cause, or whose supply or disposal produces, environmental damage. There is an *extraordinarily* large list, which takes in not only noxious pesticides and herbicides, and other hazardous or damaging chemicals, paints, detergents and so forth, many of which can be avoided entirely, but also fossil fuels, factory farm products, newspapers and junk mail, water and electricity, some of which are unavoidable, others of which can of course be used in reasonable quantities. In cases, such as electricity, where reduced consumption is what is required, the technological factor enters. So far as feasible (consistent with durability, repairability, etc.), appropriate fuels and energy-efficient devices should be selected;

- purchasing and investing along environmentally sound lines. There are very many items that people purchase, including poor people, that they do not really need or for which there are preferable substitutes which do not generate the same environmental problems, e.g. highly packaged products. For the richer especially, who have some leeway in their choices, investment and larger purchases especially, should be ethically, and so environmentally, directed. For example, funds should not be entrusted to banks or insurance companies who invest in environmentally damaging practices. For improved consumer and investment practices, a good deal more information needs to be available or circulated, both from suppliers and by engaged environmental and consumer organizations, to enable people to make better choices easily. Environmentally improved investment, for example, would certainly be assisted by the rise – with the removal of state blockages to such enterprise – of environmental banks and credit unions, environmental investment funds, and so on;

- boycotting goods from persistently polluting companies or environmentally irresponsible countries; goods that are obtained by environmentally or culturally damaging methods such as some tropical timbers and hamburger meats; that are ruthlessly tested on animals; that adversely effect endangered species or ecosystems; that cause serious disposal problems, and so on;

- voting against candidates who advocate or less vocally support environmentally unsound growth or development, or who represent orthodox parties that are so committed, so far as the voting structure reasonably permits.[2]

There are many facets to types of greener lifestyles. Lifestyles which can be rich and variable. There is, naturally, a plurality of such lifestyles (perhaps matching a plurality of human types or 'natures'). They can, for instance, range from more individualistic to highly communal. Generally, however, those lifestyles will involve much reduced consumption of non-basics; they will tread and rest more gently upon the Earth; they will be less driven, and more contemplative, meditative or devoted to fruitful leisure; they will engage in less empty travel, or waiting about, devoid of green purpose, and spend more time living in place. But often those lifestyles will involve much more.

The boundary between what is environmentally obligatory and what is supererogatory has yet to be well defined (that is yet another ethical development to be attempted sometime). What is evident is that there is much environmental action of a supererogatory kind that deserves encouragement. Such action includes

- encouraging and assisting the growth of environmental awareness and concern in others, especially among close associates and students, and discouraging indulgence in anti-environmental activity and for anti-environmental spectacles;

- direct non-violent action against environmental destruction;

- joining environmental groups and participating in their efforts;

- participating in alternatives to mainstream economic activity, such as green banks, bartering circles, and so forth.

It also includes, a variety of action of a rather different cast, such as

- organic gardening and correspondingly sound conservation farming;

- environmentally sound tree planting (e.g. *not* coniferous or eucalypt monocultures);

- sorting waste, recycling and reusing or composting materials, where feasible (e.g. not where the energy costs of recycling are excessive);

- avoiding highly processed and heavily packaged foodstuffs;

- avoiding throwaway goods or packaging;

- repairing durable goods, which are shared;

- reaching and influencing others, through speaking out, educating, letterboxing, and so on;

- becoming elected to a council or organization on a green basis or ticket.

These lists are but illustrative. There is much other worthwhile practical action that a dedicated individual, family or household can undertake (many detailed examples are supplied in texts like Seymour and Girardet, though like the run of texts they shy away from even mentioning what is critical, concerted political action).

There is a sensitive question as to how far individual and small group action against the forces of environmental degradation and vandalism should be taken. Practices such as civil disobedience cause no serious ethical issue; of course principled ethical resistance to bad laws can be justified. But to what extent is it permissible to engage in ecodefence tactics (set out in detail in *Ecodefense*), which include tree spiking, road destruction, and sabotage of environmentally destructive machinery? Ecodefence tactics are attractive, especially in individualistic cultures, because using them an individual (alone even) can *do* things that make an evident difference; and they are fast, they can yield immediate results in situations where time is of the essence; but they are highly controversial. Ecodefence groups are seen, generally unfavourably, as a mix of 'Green Luddites' and 'Green Guerillas' (neither comparison is very satisfactory; the latter description is inept, given that they generally avoid violence, in ordinary senses, and intend none, carry no weapons, and only venture out on occasional forays). One reason for the controversialness of the tactics is that ecodefence activities conflict, through direct commission, with the laws of the state, which are commonly arranged to allow, or even encourage, environmental devastation and vandalism. We should want to argue that, for state laws to deserve respect, they should enjoy rational bases and at bottom ethical foundations. There are, however, other more pragmatic reasons for eschewing ecodefence strategies; namely that (like violence, with which they are wrongly identified, which is mostly inadmissible) they tend to alienate an otherwise increasingly supportive public. For this sort of reason, ecodefence tactics, largely worked out in USA, have not been much adopted, or gained endorsement, in Australia.

SPECIAL PEOPLE AND GROUPS; MEDIA, EDUCATION AND UNION; MAKING EVERY ONE SPECIAL

On many issues individuals or households, proceeding on their own, can make little difference (unless perhaps they resort to disobedience or ecodefence strategies). But they can provide important *examples*, opening the way for others to follow, or motivating or encouraging others. Sometimes individuals can of course make a substantial difference; for instance, they are in a position to take legal action against a polluter or other vandal. Very differently, there are charismatic figures, with a following, who have environmental commitments or alternatively acquire them. Individuals can be much influenced by the conversion of high-profile or key players to a more environmentally aware and concerned position. If key influencers advocate environmentally sound positions, then their followers or fans will be more receptive to messages about environmental ethics and environmental quality. Key influencers may be found at all levels from the neighbourhood to the nation-state.

211

A 'Green Gandhi' could go a long way in helping to implant an environmental ethic. Francisco (Chico) Mendes, a Brazilian rubber tapper who spoke out and organized rubber tappers to save the rainforests, was close to a 'Green Gandhi', and he met the same fate – assassination. He attempted to attain "the prohibition of any environmentally harmful activities in the area, and the suspension of official financing and incentives for such activities ... in order to slow down the destruction of the Amazon".[3] Yet, like Gandhi, his example and inspiration live on. Hopefully new 'Green Gandhis' will emerge, both in Amazonia and elsewhere. Presently the Earth needs a whole cohort of them, well distributed through environmentally sensitive regions, as well as clusters of high-profile environmental custodians. There are many vacant niches to fill; many individuals can become environmentally special and significant, should they choose.

There are certain individuals, who can make an impact, given the support system they already feature in: for instance, media personalities, entertainers, public lecturers, politicians. Except for politicians and economists, business columnists and financial commentators, an increasing number of these elevated individuals are already environmentally on side to some extent. But many of them do not carry, or have an opportunity to present, appropriate ethical messages. It is important, then, to generate opportunities. There are many ways, this can be done: at clubs, at meetings, at demonstrations, through social gatherings or events, such as theatre, carnivals, festivals, and doings generally.

Thus, for instance, reorient social events in ways that connect them with environmental concerns. Connect carnivals in Rio de Janeiro with the destruction of rainforests along the Amazon. The environment is an integral part of many cultures. The connection between the destruction of the environment and the destruction of culture must be made and made strongly. The suggestion that the destruction of one environment and one culture is necessary to the development of another holds no water. To destroy an Amazonian Indian culture and the rainforest surrounding the people of that culture does not really promote the American or non-Indian Brazilian cultures. Brazil is the poorer for the loss of both an ancient way and a section of rapidly dwindling rainforest. And the Americans have lost yet another place to send a phalanx of anthropologists. Like the 'Laboratory Argument' for the preservation of the environment, the preservation of another culture saves an anthropological laboratory, to slip into shallow argumentation.

Closely associated with the environmental reorientation of doings, missions and so on, are subcultural movements, such as 'World Beat' in music. More and more musical groups and musicians from countries such as the United States and Britain are going to countries such as Brazil and South Africa for musical inspiration and returning to their own country to make a political as well as a musical statement. Examples of this are Paul Simon's *Gracelands* album recorded in South Africa and Sting's tour of Brazil. These musicians present to the world the musical influences of the countries they visit and make the world aware of their cultural richness. The same can be done with the environmental richness. Cultural events can present messages of an environmental ethic as well as entertainment. Many songs already exist that give an environmental message. One of the best known examples is Joani Mitchell's lyric, "They'd pave paradise to put up a parking lot".

Not so far removed from entertainment, and using much the same media and methods, is education. Not usually public figures, but very important in their influence, are teachers (and different in their roles, parents and peer groups). While some teachers are accomplishing fine work on environmental issues, not enough elements of underlying environmental ethics are getting through. So there is a need for reaching the teachers, bringing more teachers in, and teaching the teachers about environmental ethics, and its importance, and some of its social and political ramifications. But presently there is a severe lack of requisite teaching infrastructure, for instance, suitable teaching texts and teaching kits on deeper environmental ethics. The production of such teaching material is a matter of some urgency. If environmental issues are considered at all, they are generally tacked on to courses such as biology, geography (where environmental studies often survive) and social sciences; they should be made an integral part of these courses, and they should also feature in courses such as history (e.g. as biohistory), economics (eco-economics), and so on. None of these more conventional sorts of courses provide much scope for environmental ethics; that too should change. But newer introductions (in Australia) such as 'philosophy for children' do; and philosophical discussion of environmental problems of a deep character should be worked into these exercises.

Education, and information availability, matter, not merely for teachers and those they teach, but much more widely, to consumers, to parents, to managers, to workers, to everyone. In many parts of the world, eastern Europe for one major region, information concerning environmental problems is very restricted; workers are unaware, for instance, of the real effects of pollution on themselves, upon other creatures, upon forests and lakes. Indeed there are many residents of the heavily polluted city of Budapest who are not aware that the local air is polluted. Even in many Western countries, where information is rather more freely available or accessible (though much remains to be achieved on information freedom fronts), companies are often inclined to withhold information on known or probable effects of their products on health and environments. There is real need, especially in eastern Europe and Asia, for a 'green glasnost'. The associated phrase 'green perestroika' has been applied by some Eastern Bloc environmentalists, to refer not so much to green restructuring of enterprise – a much overdue desideratum, given the environmentally-unfriendly character of much state socialist industry – as to the development of environmental awareness. "Apart from trying to avoid persecution from officials, the biggest challenge facing Eastern Bloc environmentalists is public ignorance".[4] Environmentalists in CIS recognize that a major problem is that of raising "'environmental consciousness' which ... 'should not only be the knowledge of prominent scientists and writers, but should become the property of the whole society'. It is for this reason that Greenway – the organization which tries to co-ordinate environmental activities throughout the Eastern Bloc – regards one of its primary functions as being educational. Much of its effort is put into organising training courses and public lectures".[5] Similar efforts are needed in *most* of the world, particularly the diverse Latin, Islamic and Asian worlds.

To circulate information, advantage should be taken of increased coverage of environmental happenings, mostly problems or disasters, presented in newspapers, periodicals and other media channels. In some

214

pockets of the West it is a matter of expanding, improving and deepening the messages of media that are already there. Have sympathetic commentators explain the underlying philosophical concerns of releasing two whales trapped in the Bering Sea. Inform, and if need be indoctrinate, media personnel about environmental ethics. As already observed, as a form of educational communication, good television and radio programs with the right messages would be an excellent way to present elements of environment ethics. But effort *is* required to expand and deepen most media presentations. It is suggested that newspapers could run, instead of the occasional environmental item or column, environmental sections after the fashion of financial sections, which would feature, for instance, in-depth studies of particular issues or places of concern, and would include columns on prominent (or demoted) environmentalists, and so forth.

Not only should the news functions of the media be exploited, but the advertising functions as well. Advertisers could be encouraged to promote products that impact more lightly on the environment and to use environmental aspects honestly in their advertising. Furthermore, they could discourage the purchase and use of environmentally unsound or unnecessary products. People wearing fur coats could be made laughing stocks instead of being promoted as glamorous; young men equipped with rhinoceros horn daggers or swords could be presented with derision or as sexual failures. Then show the rows of dirty factories standing behind the elegant shop-front display or the superclean executive and other beautiful people.

Teachers form but one union, already moderately well disposed, that it is important to reach and, for the most part, to educate further (of course it will be teachers teaching the teachers). There are other unions to reach, some to try to redirect, some, with a record of environmental action, to encourage to continue. In the 1960s and 1970s certain trade unions in Australia, notably the Builders Labourers, started 'Green Bans' to preserve buildings of historical significance and to preserve green belts as recreational areas within cities such as Sydney. Besides preserving significant urban elements, of the built as well as the natural environment, this movement demonstrated decisively (what should not have

215

been in doubt) that environmental concern is not simply the idle pastime of middle-class dilettantes. The 'Green Ban' movement (export of which other states sharply curtailed) was eventually broken by combined stratagems of developers and states; it should be reconstituted on a broader union basis. More recently, trade unions in Germany and Sweden have become concerned with issues such as acid rain that affect their livelihoods. These active trade unions need to be appraised of the ramifications of environmental ethics and the comfort and support they can offer. There are other groups, but weakly unionized such as smaller farmers, or actively engaged in environmental devastation, such as timber workers, where the task is more difficult, but even so not nearly as difficult as it was only a few years ago. For farmers are beginning to realize the heavy health and environmental costs of intensive business agriculture. Many smaller sawmillers have already been closed down; many of those that remain have seen that their future livelihood is threatened or will be eliminated by present overexploitative forestry practices. Some of the farmers have already joined forces with environmentalists, and joint exercises in combating soil erosion and several types of soil degradation (salinity, acidification, etc.) are now beginning. There is a growing trend, too, still very small and in need of support, away from chemical farming towards organic and more biologically integrated practice. Some sawmillers also, to the consternation of big forest industries, have linked up with environmentalists. These breakaway groups deserve support, and assistance in getting better organized, especially to resist large industry pressures; as well they would profit by further education, in environmentally sounder practices in their spheres, and as to the ethical bases of these practices. In getting requisite messages broadcast widely, it helps to link environmental ethics with everyday and workaday activities. A fundamental principle of deeper environmental ethics that must be got across is that humans do not stand above or even apart from the environment but are part of it, even if some of their everyday needs modify the environment or shelter them and even isolate them from it. So their modifications – where modification is required at all, not everywhere – should be at most low-impact modifications, for which they take responsibility, never devastation or destruction of habitat, their habitat.

LARGER GROUP INITIATIVES AND ACTION:
CHURCHES AND CORPORATIONS,
GOVERNMENTS AND MOBILIZING CHANGE

At higher levels of organization, such as those of a region or state, suggestions for initiatives and action, though sometimes needed right now, await further development of deep environmental ethics and associated theory, itself an urgent desideratum. Indeed the ethics needs both to be developed and to be made a force in the world. There are now openings for attempting the latter, for example through former custodians of morals that once (some of them) rivalled states in their power and influences, namely churches. Among other things, an extensive reinterpretation exercise by skilful insiders is called for, reinterpreting scriptures in environmentally sounder fashion. For many churches the requisite scriptural basis for such a major recasting of doctrine appears in fact to be available; and suitable reorientation has already been partially accomplished by many priests, especially in third world countries. What are missing so far, it seems, in major faiths such as Catholicism and Islam, are authoritative theologians with the skill and stature to carry such a bold interpretational enterprise off successfully.

Churches are not the only transnational organizations where environmentalization should be seriously attempted. There are already transnational companies operating which provide a sufficient illustration of what could be done (e.g. Body Shop); namely, acting as a more environmentally responsible person within a deep ethical world. More could follow suite. So could transnational and international banks, by reorienting their investment practices; for instance, financing only environmentally sound and sustainable projects, and renegotiating loans that are causing unsound environmental practices, such as much third world debt is inducing. The record of most banks, thus far, is ethically appalling. Then there are international forums and organizations, such as various UN organizations. Environmentalization of such organizations will be much more difficult however, since these organizations are formed from nation-states and share their commitments to growth, development (at best slightly constrained development, laughably described as 'sustainable development'), and so on. Even so, increasingly, voices speaking for the environment or for ethics are heard. These

217

groupings should be brought together (perhaps in a coalition with other groups against forms of domination). The amalgamation could result in a powerful voice for environmental ethics.

Where most change is required is where most power and control presently reside, excessive power. That is, with states, and in many main policy, regulatory, and executive arms of states. We have already glimpsed some of the sorts of change deep environmentalism would make, for example, as to economics, and regarding legal arrangements. Many more changes should be pursued, not least reform of taxation and excise arrangements, at all levels of government. For those levies are commonly geared to environmental profligacy, exploitation and waste. But let us consider political and policy change.

At the political level, as at other levels, there are two likely least-resistance courses of action, neither of which can be recommended: Oblomov or Technofix. One, do nothing, on the questionable assumption that the environment and ecosystems will be self-regulating and self-repairing as they have been in the past, and that the human impact, while on an order equal to that of a major glaciation, is nothing more than another phase in environmental evolution with which nature will somehow cope (utterly forgetting loss of value therewith). On this course of action an environmental ethic is not needed to keep things going: ethics is shelved in favour of evolutionary determinism. Two, and more likely, continue with the present program of band-aid environmental repair and desperate technology hoping to strike a techno-fix solution, despite inappropriate technologies being part of the problem, as well as possibly part of occasional short solutions. This course of action conforms to the conservative idea, to which there is a mass of counter-evidence, that a stronger adherence to available principles and beliefs will produce a due practical outcome. Neither course of 'action' heeds rationality principles concerning proceeding with due caution, neither will deliver urgently needed changes.

Broadly, there are two ways that real change can be accomplished, both overall and also in given reaches of policy: by evolution, by slow reform steps, and by revolution. There are evident severe difficulties with both broad alternatives. With mere reformism, the overwhelming evidence is that not nearly enough will happen in time for anything but a grossly impoverished natural environment to emerge. Here for once, time is, as lawyers insist, of the essence. For much of the world's remaining wildernesses, for most of its remaining species, it is going to be all over in the next 20 years or so. Reform, even where it occurs, never happens with that sort of speed, when the initial odds are so heavily against it. And the odds are against it. The control of most states remains in the grip of growth and development cliques, as is easily verified, and as is reflected in national newspapers. For all the green noise, the attitude to the environment remains, at the politico-economic controls, largely one of tokenism: reservation of some token parks, preservation of some members of conspicuous token species, and the like. Nor are significant elements of political change to be observed. Granted, there is a new veneer of environmentalism, because it matters to some voters, and because features of excessive industrial growth are now all too visible, but underneath little has changed: it is business as usual, more, bigger, 'better'. More and more neutralized, monoculturalized, levelled and paved over: parks to parking lots. That's all too thinkable. Thinking the unthinkable again discloses a theoretical way out.

The alternative, seemingly inevitable if enough is to be accomplished in time, is revolution, revolution in a few key states. Deep environmentalism is, in main tenor and thrust, revolutionary, not merely reformist. Its theory is revolutionary, by past intellectual standards; the practice it imposes and encourages is also revolutionary. Though prepared to use reformist procedures, it stands opposed to mere reform environmentalism. Nor is it merely in theory that deep environmentalism is revolutionary. Insofar as deep environmentalism is committed to alternative political structures and arrangements, as Deep Ecology is to bioregionalism (and Deep-green theory is to anakyrism, i.e. practices without domination), it is committed to very substantial modification of present arrangements and jurisdictional boundaries, and thus to the erosion of national sovereignty and replacement of present states. But states are most unlikely to go willingly; they will not, often they cannot, legislate themselves to oblivion. Revolutions, overturning state constitutions (mostly not documents of great ethical weight or merit), appear the only satisfactory medicine. But were the styles of historic revolutions emulated, it would be a problematic and likely nasty medicine.

219

Such revolutions are difficult to achieve satisfactorily, given the virtual monopolies of states on coercion and armed violence, and *their* readiness to use violent means. Armies are often maintained in part for such purposes. Nor, in any event, is there sufficient public support for revolutionary activity in key states where it matters most. Things have generally got to be extremely bad for people themselves, certainly not the natural environment which most of them never see in a non-degenerate form, for any action to be contemplated. But even if a revolutionary initiative were achievable, it would be wasted, and perhaps even counter-productive should a state fall the wrong way, for instance to a totalitarian far right. For there is no adequate preparation for revolution, no due guidance for the felling of a state. An environmentally appropriate revolution requires, however, extensive (and intensive) forethought and organization. Requisite organization, well-thought-through directions, plans for action, and restructuring: such features are critical. Deep environmental groups should begin to prepare, carefully and thoroughly, for revolutionary action.

A clear-cut of states appears unnecessary; prevailing anarchist opinion that states have to be removed everywhere to topple anywhere, lacks a satisfactory basis. Revolutionary change is much more important in some states than others: USA than Russia (where revolutionary circumstances obtain), Japan and Brazil than Australia or New Zealand. Except for Brazil, the reason is not nearly as much the internal environmental records of these countries though these records are not something any should be proud of – but much more for the impact their practices have environmentally beyond the proclaimed boundaries of the state. The Japanese have been, and are, especially resource greedy when it comes to other lands' resources and environmental goods and those of the remaining commons. American consumption and resource extravagance remains gross. USA and Japan exercise much influence over some of the worst and most rapidly prevailing global environmental damage – the wholesale transformation of natural environments and destruction of species. Influence which under different ideologically-committed governments, they could readily and profitably bring to bear to stem the transformation and destruction. A prominent example concerns tropical rainforests. Japanese enterprises have a direct part in the destruction or degradation of much rainforest; indeed in the Pacific region they are the chief vandals, but not the only ones. USA, which has played a significant role in logging out sizeable parts of Samoa, Philippines and Indonesia, for example, and in defoliating SE Asia, is now profiting from the destruction of the Amazon rainforest; for example, it gets meat, cheap meat, it wins favour from internal elites who are advantaged, and it gains capital, as its banks are paid interest on loans. America remains the prime power-house of the extraordinarily damaging capitalistic growth and development ideology (Japan has of course been copying that power-house, with a few refinements and superior quality control), and operates the world-wide military power designed to maintain and enforce that ideology, under guise of 'keeping the free world free' – free for American free enterprise and, with it, environmental despoliation.

In the absence of internal change, external pressures can be brought to bear on vandalistic states. The internal policies of a country can be influenced by outside regions and countries, in a variety of ways. In addition to the sorts of economic and trade pressures that are sometimes used to coerce nations into policy change, co-operation and friendly exchanges can work well. Exchanges such as a recent visit to Australia by two pandas influence both governing authorities and the general populace. In Melbourne alone, 375,000 people saw them and admission fees raised A$250,000 for the conservation of pandas in China. This sort of exchange highlights differently from pollution exchanges, that environmental issues are international issues. The panda visit was an enactment of the slogan, 'Think globally, act locally'.

Similar sorts of practices, but with a more pointed character, could be enacted through an organization like 'Environment International', a environmental form of Amnesty International. No doubt actions of Environmental International, like those of Amnesty International, would involve interference, not meddling but warranted interference, in the internal affairs of state, and thereby lead (if state sanctioned) to modification of what is said to be a basic diplomatic principle and part of state sovereignty, that of non-interference in the internal affairs of other states. But that principle (based on a mix of untenable assumptions, e.g. as to state property rights over land annexed by conquest) deserves no supine respect, but rather needs modification, for instance to a principle of no violent interference, e.g. no military or like interference. A special environment, for example, is not simply a local heritage; it is a world treasure and an attempt to destroy it for local expediency may justify external interference. For a start, extra-national pressure could be brought to bear on nations holding environmental prisoners, the rainforest of Brazil, for example. The pressure would be brought to bear not only on Brazil, but on external exploiters, such as the United States, who are making demands upon the environment within the national boundaries of other countries. Within national boundaries, national pride may be a way of selling to reluctant governments the idea of protecting their environmental wealth. Amnesty already does this. Nations should be encouraged to consider their environmental stature in the international community.

There are several opportunities for external influence. One is through ethical or environmental investments. Another is through environmentally tied gifts. Benefactors to key players, such as Brazil which is responsible for around about 10% of the carbon dioxide released into the atmosphere annually through the burning of its rainforests, could pay off part of a country's national debt in return for the preservation of sections of rainforest. "In July 1987, the US environmental group, Conservation International, negotiated to buy $650,000 worth of Bolivia's debt in return for nearly 16,000 sq km (6,000 sq miles) of rainforest".[6] While not on the international level, a related practice is currently operating in part of Australia and in the United States. In the US, The Nature Conservancy has over the past 35 years bought over a million acres and donated a large portion of them to national parks, universities and other groups that will take care of them. The Nature Conservancy is the major private conservation organization in the USA attempting to save natural things and places; it is the leading bidder in the 'sanctuary real estate market'.

Internally, within states, the reform alternative must also be pursued – though it may promise little, ever, where much is required, soon. A little is better than nothing. There are many things a state can readily achieve that private organizations cannot. Curtailing overseas immigration is a simple example, but an important one in the Australian context. For ever since migrants began to arrive in Australia, Mammon has been the main god worshipped. Migrants have all too often arrived to make money, usually as quickly as possible, either for themselves or for their families back home, and then to return whence they came. Migration has frequently been regarded as a 'cut out and get out' operation, too often at severe eventual cost to the environment. Though many migrants in fact outstayed their planned return, a similar attitude prevails among new migrants today. Many of these people, from regions where the environment is but a background and perhaps already largely despoiled, do not care for their new strange environment. But further, they undercut attempts to change things, they add a factor of conservatism in communities and work-places very congenial to degraders and developers – which is one reason why there is employer pressure for immigration: they do jobs established Australians won't, work for lower wages, they don't strike or protest, but will help break protests by continuing working, and of course they expand retail markets, increase consumption, bolster property markets, and so on. While past migration has undoubtedly contributed richly in anthropocentric ways and has brought many benefits, contributions to greening Australia and greening Australian culture do not stand out among them: quite the reverse.

ENDING DOMINATION AND THE LONG WAR AGAINST NATURE

Humans comprise the main enemy of Nature. Even if they are no longer at war, because Nature is defeated and in dishevelled retreat, even if many humans are now well disposed, still too many humans, both jointly and sometimes separately, inevitably makes an excessive impact upon the environment. While there are a great many things, of first environmental significance, that states *should* be attempting, and environmental reformers should be attempting to have done, in most places population reduction ranks high on the list. That includes the developed world. Americans wonder why so much of their farm land is being paved over, for developments such as shopping malls. A major factor is human population increase; while the percentage increase *may* look respectably small, the gross numbers are gross: some millions each year. And the consumption of one American is, on average, excessively large. Technology might well be applied to reduce this somewhat; but there are limits to what it can achieve, and it is not being applied.

Each element in the human impact equation – number of humans × their consumption × their technology = human environmental impact – can in principle be varied to reduce environmental devastation. Yet, standard ethics have little to say on any of them (except utilitarianism, which says the wrong things, notably on the first). The main thrust of ethics has in the past been to determine what is appropriate behaviour by one human towards another. It is becoming more obvious that control of the global human population is an issue that effects not only the quality of the environment but the very quality of the existence of every human.

Leaders in countries such as Kenya would not readily accept any proposal, ethical or otherwise, that called for zero or negative population growth. Even Australia, which has achieved a below replacement birth rate, is contemplating large migrant intakes to increase the population, despite the environmental and other evidence condemning such a move, and increasing opposition to immigration. Nonetheless, a smaller human population would be a major step in the right direction. States like Kenya should be brought around, by reason and persuasion. For example, the environmental case could be argued in conjunction with a case from a related area, namely health. Without directly linking environmental protection to health issues (such as the threat of resilient diseases and AIDS), arguments could be developed for taking action to reduce or level off the human population, given that the opportunity to do so now arises in new ways. Any decent program or policy or ethic that advocated the limitation of population growth would be an ally of the environment. This is not to be understood in any way to be an advocacy of cultural genocide, merely the regulation of environmentally unsustainable growth. Any environmental ethic is after all an ethic, and ethics are not intended to harm, but to help secure well-being.

But, short of catastrophe, population reduction is a slow long-term affair. Meanwhile, much could be achieved by conservation coupled with updated technology. For example, a relatively small decline in the *increase* of energy consumption would help considerably in staving off the greenhouse effect. To achieve widespread conservation, and application of improved technology, much coordination, often at state, regional or federal levels, will be required. The same applies to many other environmental problems: from pollution reduction, through soil erosion and degradation, to trade in endangered wildlife. The control and regulations operating for environmental protection are seriously inadequate in virtually all states; expenditure on protection is minimal. Moreover on the rare occasions where there is passable legislation compliance with it is not attained, for instance it is not duly enforced.

Much improved environmental protection is but part of what is required from states (while states remain, and while they dominate so much of social life). An environmentally committed state would have to adjust to much else, from educational structure, through law to economics and politics. Above all, it would aim at a restructuring that would render new environmental lifestyles at least feasible, or still better, easier to attain than consumptive ones. We have already seen what some of these adjustments would look like if deep environmentalism made due impact; macroeconomics, for instance, would substantially change in character, cost accounting would change, taking into account the overall environmental impact of and damage produced by goods and services, and so on. A state that took the environment seriously would not assign 'Environment' to a low-ranking ministry with little impact upon major decision making and constantly under scrutiny by finance or treasury; rather the reverse would be the case. Environment would rank very highly and take precedence over development, militarization, etc. There would perhaps even be – better than what is now much favoured, a Department of Sustainable Development, a Department likely to be economically captured and dominated – what Canadian geneticist and broadcaster David Suzuki has suggested, an overarching Office of the Ecosphere, nonbureaucratically organized, designed to ensure impartiality, to which the rest of government responded.

Many of the adjustments, aimed at environmental adequacy, would ideally be integrated with adjustments to meet other popular demands: those for peace, those of the underprivileged for a fair deal. Domination of Nature is but one form of domination that past states have sanctioned and encouraged. A broad change in social arrangements would aim concurrently to remove and redress other associated forms of

225

unwarranted domination: of indigenous peoples, of subject people, of the poor or homeless, of women, of powerless foreigners, of animals. A genuine green environmental ethic stands against all such chauvinism, against all such forms of domination and oppression. Let ethics be greened!

Notes

227

PREFACE

1 With items like ethics some pushing and stretching has to be done to fit, indeed to make sense of, economic conceptions at all. After all what is *consuming* an ethic; but then what is consuming street lighting or a forest park? Certainly one can make *use* of an ethic (acquire it, take advantage of it, adhere to it, even discard it, etc.) Like theories more generally, *where* an ethics fits into economic accounts of public goods depends on how it is conceived, presented, packaged. For a theory is at bottom an abstraction, a structure of propositions which can be given various material presentations, though an aggregation of token sentences, which may be encapsulated in hard copy (books, etc) or soft copy (tapes, etc), or may be presented through a public medium such as broadcasting.

   In the latter case at least, a theory exhibits *non-rival consumption*, that is one person's consumption, or use, of the item does not reduce its availability to anyone else (though its 'availability' even to humans, may be sharply limited because it is hard to grasp). In more permanent book presentation, by contrast, whether there is non-rival use depends both upon how the book is used (normally it will survive a reading) and upon whether the book itself is a public good (e.g. held in a public library) or a private one. For the abstract conception of a theory, however, the notion of non-rival consumption hardly makes sense, because such an abstraction cannot sensibly be used. How then can we assert that an ethics or like theory is a public good?

   An ethics or like theory is a public good by virtue then of certain of its material presentations. It has material presentations, themselves also accounted ethics, which satisfy the basic definition. Namely a public good is a commodity or service or, more generally a useable, "which if supplied to one person can be made available to others at no extra cost" (Greenwald 1982, 352); the modal 'can' is essential, because availability depends on the right sort of material presentation. Because of the feasibility of other presentations, which can be privatized (as books can), an ethic is a *mixed* public good. (It is not pure because it fails to satisfy the esoteric property of *non-excludability*, "that is, if the good is provided the producer is unable

227

to prevent anyone from consuming it" – by legitimate methods only no doubt.)

Interestingly, much of what is claimed, in economic works, of public goods is obviously false of useables like ethics. For example, "the provision of a public good is a matter of *collective choice*" (Greenwald 1982, 353). But this sort of claim fails even for more normal public goods, which can be individually supplied, both by public benefactors (or malefactors if the useable does damage, is of negative utility) or by others. The problems here are not ours: they are, at bottom, that the economic theory of public goods remains in unsatisfactory shape, as the case with ethics shows.

2 Among the many examples, one of the more outrageous is Nash 1989, subtitled "A history of environmental ethics". Unfortunately the grand history Nash offers of *American* environmental ethics omits from the subtitle the necessary qualifying term 'American' (nowhere else, bar perhaps France, would it be imagined that the country comprehends virtually the whole world; for that matter, in no other region would *rights* be quite so dominant). It strikes us, by the way, that the background operational picture of Nash's work is seriously flawed, being premised upon a substantial misconception: that self-interest is and was fundamental. Proto-philosophy did not invariably begin from the notion of an isolated (selfish) self; getting such a (theoretical) notion of self established and instilled required much intellectual effort and much indoctrination. A brief survey of Australian environmental philosophy, up to 1980, is given in Mannison et al. 1980.

3 The bogus near-exhaustiveness Fox claims (in his Preface) for his 1990 text as regards Deep Ecology is only achieved to the very limited extent that it is (for much of importance does not appear), by a highly idiosyncratic and narrow selection of Deep Ecologists. (These Deep Ecologists are rather fond of claims to completeness and totality.)

Main consideration of recent Australian books, such as Fox's 1990, now appears in the sequel to this text, namely *Green Ethical Fields*, as tentatively entitled.

CHAPTER 1

1 Leibniz 1846, 33.
2 Schweitzer 1933, 188.
3 A striking recent example is Singer 1991.
4 John Langford, on ABC radio, 22 February 1993.
5 Engel and Engel 1990, 6.

6   These action basics can in turn be explained within process theory, see Sylvan 1992. A classification of agents can be drawn from there.

7   In logical terms it is defined through attribute abstraction.

8   Greek *ektos*, outside. We include the inside with the expansion which incorporates the outside.

9   Thus ethics include what it has to include, as expanded environmentally: philosophy of ecomorality, and therewith of morality.

10 A *system* is a relational structure of a certain sort. An ethic is a *propositional* system, since it has to include principles and the like; thus it involves a theory of strict logical sort. For an illustration of the use of systems formulations in ethics, see e.g. Routley 1973.

11 These objections have already been addressed in other places, e.g. Routley and Routley 1979.

12 Gunn 1986, 3.

13 Drengson 1989, 15.

14 Elliot in Singer 1991, 284.

15 That is, fashionable contemporary practice with respect to philosophy is transferred to parts of environmental philosophy (e.g. by EcoWittgenstein and AppliedQuine). Environmental philosophy may in no way interfere with industry and business. To an astonishing extent American academic environmental ethics tends to conform to such ridiculous strictures (often encouraged by a fawning pragmatism).

16 It need not however take (what may be meritorious) a 'bleeding heart' approach. There are further distinctions here, between ways of advancing causes.

17 Compare also Naess (in Engel and Engel 1990) on utilitarian approximations to Deep Ecology, made by assuming availability of enough agents holding deep ecological values or, more or less equivalently, having deep ecological consciousness.

18 Although we are explicating terms that are now in widespread usage, in a way substantially faithful to that usage (but not some of its dilute or corrupt expansions), we shall regard the terms as quasi-technical. We see these terms, as the discerning reader also may, as spelt not with a double 'e' but a triple 'e', as 'green and 'greening' (insert hyphens as need be for appropriate pronunciation, e.g. 'gre-een'). The effect is to peel off dilution and corruption of 'green' (e.g. so to include every politician, at least when any confronts concerned constituents) and to discard all those unwanted and often undesirable senses the colour term *green* has accumulated.

19 In a study entitled "Green packages in Australia", Greens were picked out from the whole sample of respondents to the Australian Election Study of

1990 in terms of answer to the question: How likely are you to join groups campaigning to protect the environment, or are you already a member? Permitted answers: (1) I am a member, (2) I am not a member but I have considered joining, (3) I am not a member and I haven't considered joining, and (4) I certainly would not consider joining. Those giving answer (2), perhaps token greens, were accounted green.

20 Thus Singer, who in the *Companion to Ethics* 1991 (on p.xiii), proceeds to present environmental ethics as a further applied ethics, alongside other ethical fields. Against the applied idea, see Sylvan 1993.

21 There is a long history, yet to be satisfactorily documented, of these sporadic attempts. What such a history will probably reveal is that much of what is considered highly contemporary in environmental ethics, had been proposed earlier, but generally ignored.

22 For a more detailed, though inevitably still schematic, explanation of this sort, see Nash 1989, 4-9.

23 This is one of Nash's example of 'the expanding concept of rights' (figure 2, p.7), revealing both a confusion of legal with moral rights, and American parochialism.

24 Schweitzer 1933, 188.

25 Reorientation of the picture, of circles within an ethical scope setting, so as to base the circles on relevant ethical features, will lead to the annular theory subsequently advanced.

26 Nash, p. 69, p. 68 rearranged.

27 *Right* has become an important player in the ethical sphere because of its modern political clout. It is a clout that is not entirely deserved, especially as rights are often simply invoked without satisfactory justification; many of those who invoke new rights wish to obtain for free what requires justificatory work. If it were not for this clout, it would be much easier to resolve 'problems' over *what* holds rights, through distinguishing *types* of entitlements: entitlements of what can itself petition, entitlements of what has certain sorts of interests (an interrelated elastic notion), and so on.

28 For a much fuller discussion of Last Person examples, see Routley and Routley 1980a.

29 On such a scheme, see Edwards 1982. Independent applied ethics centres fall into a similar hole, in consequence serving up work that is philosophically uninformed.

30 This is now carried out in the sequel to this work.

31 Initially we advanced this claim in the context of a study (Sylvan 1990) of McCloskey's work on environmental ethics and politics: that in McCloskey's text, as in that of his main model, Passmore's establishment text, there is

no well-organized classification of environmental problems, only a rather ramshackle list. But it quickly became apparent that these problems of a classification of problems are much more extensive , that it is not merely a local deficiency.

32 A touch of pseudo-Heideggerian investigation will synthesize our topics, and the stock variations, we use heavily, of *environmental* to *green*. Observe that *en-vir-on* discloses two predominant parts *on* and *vir*, that is tracking back to the classics *on* and *viridus*, or in the vernacular, *on green*. Observe further that *green* itself discloses to *gre-en*, that is redirecting the *en* or *on* and reorganizing, *on gre-en*. Thus *green* and *environ* and its complexifications are pseudo-etymological transformations of one another. *Green* is an impeccable replacement for *environment*. Disregarding the Irish, new chums, inexperience, and all that! Which *gre-een* does.

33 Concise Oxford Dictionary, entry for 'ecology'.

34 A segment overimpressed with social ecology: see especially works by Bookchin 1990 and Dobson 1990.

35 For a more detailed account of the logic of procedures and processes, see Sylvan 1992. For the most part, logics for environmental theory are not well developed. There is some work, mostly from Eastern Europe and now fairly old, on the logic of problems, but it contains little on (what is essential) the internal structure of problems (rather it is primarily propositional, including an interesting intuitionistic 'logic of problems'). Again there is some logical work, for instance by Niven and drawing again on Polish logic, on environments of animals and plants, but it remains in initial stages. There is much to be done: ecological logic is a virtually green field.

36 Technocratic ideologies tend, erroneously, to assume that all problems have solutions. Less arrogantly, we have found it asserted, without any requisite argument, that while not all problems have solutions, all environmental problems do – no doubt a very convenient doctrine. We claim to have demonstrated the contrary: that some environmental problems are classically unsolvable. These issues will recur below, especially in the sequel, in connection with managerialism and techno-fix.

37 Values enter again into rankings of impacts negatively or positively, as to weight or intensity of impacts, and so on; these are commonly evaluative matters. Accordingly too, values enter into what counts as a problem, how severe it is, and similar. It is with this that an important cross-classification of attitudes to the environment connects: the division into shallow, intermediate and deep, in terms of which much ensuing discussion is structured.

38 In the usual ontologically-prejudiced jargon, it makes only an existential claim, not a constructive one. As to functional character, no more is asserted therein than that where factors *F1* and *F2* coincide, i.e. *F1* = *F2*, then with *P1* = *g (F1)* and *P2* = *g (F2)* for suitable *g*, *P1* = *P2* - for any of the factors included. For presentation of such production functions, see standard economic texts.

39 There are major deficiencies both in standard factors of production, which have been stigmatized as 'hopelessly heterogeneous aggregates' (e.g. land is said to be a heterogeneous aggregate of physical structures, namely staging and storage spaces, materials, including stocks and flows, and energy potential) – and with proposed replacements, typically drawn from heavily reductive technological perspectives of physicalist or thermodynamic cast. In short, then, the theory of production, from the factors up, stands in need of overhaul, of non-reductive environmental kind. Here is one of several points of entry for a genuinely gre-een economics.

40 For a practical assessment of the considerable life-cycle impact of washing machines, see Crothers 1993.

41 For one such sectional breakdown, see subsequent investigation of sustainable development in the sequel, for which the present discussion prepares some ground.

42 Of course the controversial equations, from the operation of which the still more controversial larger term catastrophic collapse of critical parameters followed, are still quite some way off. 'Limits to growth' apparatus is important, however, for impact evaluation, even if no such limits emerge, so rapidly or even at all. But there is less and less room for supposing that human civilization will manage to slip by terrestrial environmental limitations.

43 See further Sylvan 1994, chapter 9.

44 In the classification that follows we make heavy use of Wells et al., a course notebook which provides further information on several of the theories considered and gives lists of proponents.

45 The affluence theory is exemplified not only by Rivers 1977, but by Trainer 1985 (of *Abandon Affluence!* fame) and by many grass roots environmental groups. We separate out social ecology, which some sources lump under affluence, as a CT theory.

46 This sort of theory also is promoted by many proselytising Americans, e.g., recently by Lovins 1987.

47 Thus, for example, we do not accept White's 1967 findings, about *the* historical roots of environmental problems in Christianity. Christianity, though in several respects an improvement on tribalism and paganism,

simply accentuated certain features, such as human chauvinism, already represented in abundance in predecessors. While no doubt, Protestant ethics, reflecting a particularly exploitative form of Christianity, had, as a significant matter of historical contingency, a leading role in the rise of modern industrialism, many contemporary environmental excesses cannot be satisfactorily sheeted home to Protestantism, but indict other PCT supporting ideologies. For more on this issue of roots see Sylvan, 1993a.

48 The comparative figures for paid-up members of environmental groups come from the same period (recession may have altered comparisons): see Chubb 1988, 16.

49 Such groups egoism is critically examined in Routley and Routley 1979, and its initial appeal removed. The criticism there made also tells against this high redefinition move. As we have already indicated, a pervasive philosophical (and intellectual) error is that of unnotified *redefinition*. Critical expressions are lifted out of their normal uses (which dictionaries try to record) and adjusted, redefined, in high or low ways. The term *crisis* has been shunted around in some amazing ways.

50 Brennan 1991, 279.

51 Gever 1991, xliii; the language of crisis which runs through the book enters too on this page.

52 Meadows 1992, xiii, reiterating assertions of their earlier *The Limits To Growth*.

53 See especially p. 226.

TWO DECADES OF ENVIRONMENTAL ETHICS

1 Austin 1989, 42. Frankly scepticism appears an appropriate attitude concerning such themes. If, for instance, humans were simply to withdraw, and lay off, many environments would eventually heal themselves (though species impoverished). The theme shares some of the implausibility of its analogues, as variously (but inconsistently) restricted to science, economics, business, or like highly approved activities. Thus, for example, only scientists can show us the way out of the mess applied science has helped to produce – so some scientists fondly imagine.

2 The relationship has long been schizophrenic. The backdrop view considered in the text is a substantially urban one. Outside of the sheltered cities, the relation has often been one of war. But here too there are reasons, different reasons, for reassessing human-nature relations. For the long-running, but absurd, war against the environment appears substantially

over (except in a few frontier areas such as Brazil). Nature, although still a force to be sometimes reckoned with, although still occasionally marshalling impressive forces, is largely conquered.

3   Frankena 1979, 3.
4   Routley 1973, 206.

CHAPTER 2

1   Passmore 1974, 91.
2   Passmore 1974, 178.
3   William Grey was formerly William Godfrey-Smith. See Godfrey-Smith, 1979.
4   Passmore 1974, 102.
5   Passmore 1974, 102.
6   Godfrey-Smith 1980, 31.
7   Austin 1988, 48.
8   Godfrey-Smith 1979, 311.
9   The nice arguments concerned, several based on investigation of island habitats, are marshalled in Wilson 1988, introduction.
10   Passmore 1974, 107.
11   Passmore 1974, 56.
12   McCloskey 1983, 34.
13   Passmore 1974, 124.
14   Passmore 1974, 124.
15   Midgley 1975, 110.
16   Although these damaging traditions are regularly referred to as 'Stoic-Christian', by Passmore and many others, there is really little textual basis for attributing to Stoics the picture of, and attitudes to, *nature* involved.
17   Passmore 1975, 253.
18   Passmore 1975, 254.
19   Val Plumwood was formerly Val Routley. See Routley 1975, 172.
20   Passmore 1974, 24.
21   Passmore 1975, 259.
22   Passmore 1974, 28.
23   Routley 1975, 174.
24   Passmore 1974, 111.
25   Passmore 1974, 124.
26   Passmore 1974, 124.
27   Passmore 1974, 124.
28   Passmore 1974, 175.

29   Passmore 1974, 175.
30   Passmore 1974, 175.
31   Passmore 1974, 178.
32   Routley 1975, 180.
33   Routley 1975, 184.
34   Attfield 1983, 5.
35   Attfield 1983, 6.

CHAPTER 3

1   Leopold 1966, 262.
2   Leopold 1966, 219.
3   Passmore 1974, 116.
4   Callicott 1982, 164.
5   Godfrey-Smith 1980, 33-34.
6   Godfrey-Smith 1980, 34.
7   Rachels 1976, 223.
8   Callicott 1979, 76.
9   Rodman 1977, 95.
10   Leopold 1966, 219-20.
11   Leopold 1953, 153.
12   Rodman 1977, 87.
13   Townend 1981, 8.
14   Narveson 1977, 164.
15   Bentham 1823, 283.
16   Singer 1977, 26.
17   Singer 1989, 37.
18   Singer 1989, 37.
19   As regards population, see Sylvan 1991.
20   Benn 1975, 9.
21   McCloskey 1965, 126.
22   Singer 1977, 246; 1979, 253.
23   Singer 1976, 150.
24   Singer 1977, 24.
25   Singer 1976, 152.
26   Singer 1979, 195-6.
27   Singer 1979, 196.
28   Devall and Sessions 1985, 55.
29   Abrahamson 1985, 25.

CHAPTER 4

1   Benn 1975, 5.
2   Devall and Sessions 1985, 226.
3   Sylvan 1985, 43.
4   "Apron" figure after Naess 1983.
5   Naess 1973, 95-98.
6   Naess and Sessions 1984, 5. Amusingly these substantially disjoint formulations from 1973 and 1989 appear juxtaposed on opposite pages (28 and 29) of Naess 1983, where Naess also asserts that "platform formulations are ... supposed ... to express the most general and basic views [supporters] have in common"!
7   Sylvan 1985, 52.
8   Notably Fox, see esp. 1990. Also Rothenberg, e.g. introduction to Naess and Rothenberg 1989.
9   For the club, see Appendix A of Fox 1990.
10 We have already made two detailed critiques of Deep Ecology and its variants (and plan more): see Sylvan 1985 and "A critique of (wild) west deep ecology" in Sylvan 1990. We do not try to summarize that material here. Rather what we attempt is a different *traverse*, which enables a softer view of some previously glimpsed territory and offers a perspective on some new or previously unexamined Deep Ecological land forms.
11 Naess 1973, 96; Naess and Rothenberg 1989, 28 (order inverted).
12 Naess and Sessions 1984, 5.
13 Sylvan 1985, 5.
14 Naess and Sessions 1984, 5.
15 A slogan Naess now has definite reservations about: see Naess and Rothenberg 1989, 166ff.
16 Naess and Sessions 1984, 6.
17 Regan 1982, 199. Naess and Sessions hold intrinsic value is synonymous with inherent value.
18 This nice try at repairing egalitarianism – distancing it from equalitarianism – borrows from deep-green theory.
19 Naess 1973, 96; Naess and Rothenberg 1989, 28.
20 On these obligations and duties, which Naess tries to render compatible with biospheric egalitarian by insertion of the term 'in practice', see Naess and Rothenberg 1989, 170.
21 Singer 1977, 24. Given Singer's problems and his rather unconvincing resolution, his approach does not really afford a successful working model.
22 Sessions 1980, 398.

23 Such weakening is a methodological practice that has been applied to virtually all the difficult or ecologically effective principles of Deep Ecology: the metaprinciple is one of retreat towards ecological platitudes.

24 Fox 1990, 223-4, Sessions' view is quoted in Fox.

25 *Ibid.*

26 Naess and Sessions 1984, 6; parenthetical remark added.

27 Thereby revealing the incoherence of non-atomistic egalitarianism. On this logical incoherence, see the criticism of biospheric egalitarianism in Sylvan 1985.

28 Under early Deep Ecology the tension would be reduced by insisting that such part/whole talk was only at a superficial level of communication (see below). There is another more satisfactory, less linguistic-conceptual way of restoring coherence given this metaphysical shambles, the route deep pluralism takes (on which see Sylvan 1994). According to this metaphysical pluralism, while the ordinary natural world consists, in the commonsense way of wholes and parts, there is an underlying whole (of which Deep Ecologists may have a Gestalt) which has no parts, which answers to the demands of total holism.

29 Fox 1984, 196. Fox too has moved on; anti-dualism as the central intuition has been superseded by self-realization as the central ecophilosophical idea.

30 Naess 1973, 95.

31 Naess 1973, 95; Naess and Rothenberg 1989, 28. For a fuller explanation and criticism of total holism, see Routley and Routley 1980b.

   Though we have taken advantage of several swift slides in the Deep Ecological literature, there are significantly different issues to disentangle here: not only the status of curious sorts of wholes (and Gestalts), and their various relations to parts, but the structures of things and the role of (various sort of internal) relations in their characterizations, the rights and wrongs of dualism, and so on. While these metaphysical exercise are far from ecologically irrelevant, they exceed our present brief.

32 Thus, for instance, Naess's important widely distributed and reprinted 1983 article.

33 Naess and Rothenberg 1989, 165; rearranged. For what ecosophy is alleged to do, see 164.

34 Devall 1980, 310.

35 Continuing the quotation of Naess from Fox 1990, 224, on what biospheric egalitarianism now comes down to, is its pathetic reduced form.

36 Attfield 1983, 4.

37 Reed 1988, 3.

38 Reed 1988, 3.

39 Thus Naess and Rothenberg 1989, 8; see also 173.
40 Naess 1978,4-5.
41 Devall and Sessions 1985, 67.
42 Naess n.d., 2-3. Evidently, Deep Ecological extensions of *self* resemble Land Ethic expansions of *community*, both in character and ecological purpose.
43 Taylor 1987, 100.
44 Sylvan 1985, 27.
45 Sylvan 1985, 24. For a sustained criticism of self-realization directives, see Sylvan 1990.
46 Naess and Sessions 1984, 6.
47 Margalef 1968, 11.
48 Southwick 1972, 141.
49 Southwick 1972, 237, 239.
50 Naess 1973, 96.
51 Ehrlich, Ehrlich, and Holdren 1977, 128.
52 Naess 1973, 97.
53 Sylvan 1985, 15.
54 Commission for the Future 1989, 5. Latest World Resources Institute Estimate is an average for the 1980s of nearly 17 million hectares per year deforestation (World Resources Institute 1992, 111).
55 Ahmad and Kassas 1987, 4.
56 Gould League 1982, 3.
57 Bodian 1982, 10.
58 Ovington 1978, 106.
59 Poster from the Singapore Zoological Gardens, *The Value of A Tree*, based on *Update Forestry*, Michigan State University.
60 Bennett 1986, 91-92.
61 Frankel and Soulé 1981, 26.
62 Naess 1984, 4.
63 Bodian 1982, 10.
64 Bennett and Sylvan 1988, 155-156.
65 E.g. that of Christopher Reed's mentioned earlier.
66 See Devall and Sessions 1985, 21-24.
67 Sale 1985, 43-47.
68 Sale 1985, 45.
69 Sale 1985, 45.
70 Sale 1985, 46.
71 Jim Dodge quoted in Devall and Sessions 1985, 21.
72 Reed 1988, 3.
73 Sale 1985, 50.

74 Naess and Rothenberg 1989, *Ecology Community and Lifestyle* (ECL). About 60 pages of 190 goes into two chapters on these topics. As well connected material appears in other chapters.

75 P. 162 , with rearrangement. All page references in this section are to ECL.

76 P.160.

77 Into this category falls the material on the green pole of the political triangle, p.133ff, and much of the checklist and commentary on ecopolitical issues p.135ff.

78 Though Naess has an earlier book on pluralism, pluralism makes no adequate entry in ECL.

79 See e.g. p.135 top.

80 P.131. The point about the substantial independence of production from individual, and consumer, choice is attributed to Galbraith.

81 P.132. Main elements of this astute analysis of a now familiar political predicament are attributed to Schumacher. The depth is again an imported depth, from extra-ecosophical sources.

82 It means as well that the clever choice of 'Deep Ecology' as movement *title* carries, only along with advantages of making the ideology look like some sort of deep science, a political downside: that it too may look (before proper inspection) cleanly scientific, value free and politically neutral, just as "very strong forces [are] trying always to show that questions having to do with ecology are cleanly scientific" (p.132). Then in turn science can be used, as Naess indicates, in support of elements of the dominant industrial paradigm. Governments hire experts (rent-a-professors) to report and publish findings highly compatible with industrial advantage.

83 See e.g. p.160-161.

84 P.156.

85 See the checklist of issues on p.136.

86 All quotes are from p.156. There are less than two ragged pages on these big issues.

87 This modern view of social structure, within the political economy, was enunciated centuries ago by Smith; see Roll 1992, 154.

88 Things are changing on the edges of the Deep Ecology front. If texts like those of Eckersley 1992 and McLaughlin 1993 should become accredited as Deep Ecological output, then ecopolitics in Deep Ecology will turn out to be more sophisticated and rather more interesting than ecopolitics within ecosophy, which is what we are primarily considering. However much of our criticism, especially that concerning lack of Deep Ecological social and political vision, transfers to such outlier works as well. Criticism of recent Australian productions on green political theory, a significant set,

is now reserved for the sequel (order now!).

89   For one much fuller critique of bureaucracy see Burnheim 1985.

90   P.159. Similarly p.135 where Naess adds 'the less mental change in the green direction, the more regulations', a far from universal truth.

91   Contrast p.158 with p.157. For Naess's ambivalence see also p.159.

92   Pp.141-6. For a much superior anarchist version, see Kropotkin's works. It seems to us that *these* anarchistic features (duly decoupled from shallow themes of classical anarchism) are what Deep Ecology *ought* to be advocating, clearly. It should strongly encourage a political economy specifically aimed at substantially lowering human impact on environments, through changed impact-reducing social and technological organization.

93   Cp. 145.

94   The passage borrows from just such a superficial put-down of work in environmental ethics and politics.

95   P.152, also p.106, p.110. The entertaining oblique material comprises Naess's criticism of criticisms of limits to growth investigations, pp.152-3.

96   P.110ff. While the criticisms are alright, so far as they go, basic explanation is not. A supporter would not be able to furnish an adequate account of how GNP is constructed on the basis of what Naess supplies. Naess's chapter is not part of that handbook supporters need, a handbook that is not yet available. Supporters would do better to begin with Daly and Cobb 1989 and other work by Daly (despite the lack of depth therein).

97   P.106.

98   Whence of course modern economics emerged (p.105). A long line of earlier economists, from Xenophon on, were "*ideological advocates* primarily of the people who had the property, the landowners" (p.104, itals. added). As to Naess's temptation, see p.116.

We consider Naess's radical position essentially correct. He should have succumbed to temptation!

99   Pp.115-116. There are nice points embedded here. It should be recorded that Naess's attitude to GNP and economic growth here contrasts sharply with that of some supporters, who imagine that Deep Ecology supports a steady state economy, and so on. Now, more recently, there is tension in the ranks over whether Deep Ecology endorses sustainable development; again Naess has vacillated, this time trying to have it both ways.

100 Pp. 124-125. For a fuller account of these issues, see Sagoff and (building critically on Sagoff 1988) O'Neill 1993.

101 See p.123. But here too there is ambivalence, as will appear.

102 P.107.

103 P.105.

104 P.110.

105 P.109. Naess claims to extract these hypotheses from a Scandinavian text on social economy. There is some confusion in Naess's text on this page, with the connection of (B5) and (D9) wrongly inverted. Really the section needs careful reworking.

106 P.109. continuing.

107 P.109. Naess rightly contests stock assumptions about the "good times" (p.116), calling for a major ideological change. "The ideological change is mainly that of appreciating *life quality* (dwelling in situations of inherent value) rather than adhering to an increasingly higher standard of living", in the per capita GNP sense (Naess and Sessions 1984, 5). In fact that sort of distinction is now widely appreciated, and even statistically approximated. But appreciation of the difference has produced few profound changes in economic practices or organization; again, however, there is little Deep Ecological *detail* as to what change would supposedly result, though change would presumably conform to such slogans as "Simple in means, Rich in ends"!

108 See p.116.

109 See p.118ff. esp. p.119

110 P.120. Naess imagines this latter is inevitable. But it is not. It depends on a particular, and defective, hypothetical deductive methdology, to which Naess is wedded.

111 P.122.

112 P.123.

113 Naess recommends, in that familiar inconsistent fashion, both eliminating welfare terminology and defining *welfare*. Whilst 'eliminative definitions' accomplish elimination of a sort, Naess's definition will hardly do. It is of "level of welfare as the level of agreement of actual life with a life in harmony with one's norms and values"!

114 P.143. Such argumentation is extraordinarily well established. It was presented in a very forceful way by Smith, way back in *The Wealth of Nations* 1776, vol i. p.17.

115 Here are some of the many rocks upon which comparative advantage theory founders and shatters. For others, see Routley and Routley 1973. Similar reasons are among those that underwrite proper green objections to GATT principles.

CHAPTER 5

1   Main sources of deep-green theory are Routley and Routley 1973, 1980a and 1980b and the Green Series (*Discussion Papers in Environmental Philosophy*), published from the Research School of Social Sciences, The Australian National University: see also bibliographic entries under Routley, Plumwood and Sylvan.

2   The tabulation expands Sylvan 1988, 3.

3   The detailed argument for this proposition is presented in Routley and Routley 1980a, 97-108, where chauvinism is spelled out a bit further. A truncated version of this argument appears in Fox 1990, 16-17.

4   Though the theory is so labelled in several publications, the label, which presumably should stay, is misleading, because some ethical relevant categories may overlap one another. The concentric theory (developed in Wenz 1988 and others, perhaps in steps of Leopold), which incorporates a similar mistake, can be seen as a subversion of the annular model.

5   Routley and Routley 1908a, 103.

6   Routley and Routley 1980a, 108.

7   It is interesting to note that Bob Hawke, former Australian Prime Minister, in his locally famous political statement on the environment, acknowledged the intrinsic value of nonhumans, even if he was unwilling to disconnect their value from instrumental human use. Hawke states, "Plants and animals have value in and of themselves but they are also the basis of the ecosystems which support human life" (Hawke 1989, 18).

    Significantly earlier New Zealand political arrangements had proceeded much further than rhetoric, explicitly recognizing intrinsic value of natural items and systems in legislation, a recognition which has already had practical bearings.

8   Moore 1903.

9   For development of this comparison, see Sylvan 1992a.

10 For (necessary) elaboration of this condensed exposition, see Routley and Routley 1980a and Sylvan 1986 and 1992a. Main elements of the so far scattered value core of deep-green theory can be assembled from these sources and what they refer to (that is, from two layers of reference only).

11 For technical details, which many have found perplexing, see Routley and Routley 1973a.

12 The preceding two paragraphs are drawn from Sylvan 1992a. A fuller, but earlier, exposition appears in Sylvan 1986.

13 For such an approach, see Sylvan and Bennett 1990.

14 For details of the deontic theory see esp. Routley and Plumwood 1984. For the theory of rights, agent and animal rights, individual and species rights,

see Routley and Routley 1980a, 123-5 and 149-50 and further the third essay in Sylvan 1986. On interconnections of deontic notions with the underlying value theory, see also the second essay in Sylvan 1986.

15 Routley and Routley 1980a, 104, 107, 109.
16 Routley and Routley 1980b, 319.
17 Routley and Routley 1980b, 319.
18 Routley 1982, 32.
19 Routley and Routley 1980a, 174.
20 Routley and Routley 1980b, 323.
21 Routley and Routley 1980b, 313.
22 Some earlier investigations on deeper-green economic theory, some going back before 1970 – on such topics as selective economic growth, deep benefit-cost analysis and decision theory – are included in the appendices of Routley and Routley 1973. The first substantial statement on deep-green political theory appears with Routley and Routley 1980b. It has since been enlarged, and modified, by studies in the Green Series (esp. Sylvan and Bennett 1990) and on anarchism, much of which is soon to be gathered in Sylvan 1994a. The structuring pluralistic metaphysics is presented in Sylvan 1994.
23 Much of this work too has been presented in the Green Series. And much of what has not been published there is referred to there.
24 It is not Naess's only objective. Naess has a long record as a committed pluralist, a commitment rather forgotten in Naess and Rothenberg 1989.
25 For more detail see the appendix of Sylvan 1985. Naess and others advocate pragmatic use of anthropocentric arguments compatible with deep results where they will work, with deeper considerations as fall-back. Compare the stance defended in Routley and Routley 1973.
26 Details are collected (from earlier articles) in Sylvan 1994.
27 These principles, for all their difficulties are not discarded; see Naess and Rothenberg 1989, 28.
28 For fuller explanation of the incoherence of these principles, see Sylvan 1986. Some of the attempted reformulations of the egalitarian principle are trivial, and for that reason unsatisfactory. For example, it is said (e.g. by Devall) that all the principle asserts is that humans are part of nature. Similarly (thus e.g. Bennett 1986) the principle is, as property understood, supposed to make humans part of nature, rather than either separate from it or dominant over it. But it is virtually a truism that humans are an evolutionary part of nature. And a non-domination principle (as yet unformulated, and possibly shallow) is far removed from any egalitarian principle.

29 Details are given in Routley and Routley 1980b.

30 And now, in deep pluralism, a Wholle. Interestingly, the *man-in-environment* image that Naess maligns in reaching total-field holism reappeared above, in neutralized *agent-in-environment* form, in characterizing the whole moral enterprise.

31 More or less stock logical models will serve, e.g. take 3 individual horses in a race, with predicates reflecting inequalities (e.g. in race capabilities) and consider handicap predicates (e.g. weights assigned). Of course a vacuous 'equality' can be infiltrated; e.g. no horse is favoured, so every horse is 'equally' favoured; but such a modifier merely idles.

32 For much fuller discussion of these and connected issues, and for a (different) theory of *self,* see Sylvan 1990.

33 Despite efforts based upon misconstruals, by commentators such as the abovementioned Christopher Reed.

34 On the limited extent of deep-green theoretical commitment to Enlightenment precepts, see Sylvan 1994, chapter 10, where anti-Enlightenment themes are also critically explored.

35 As exchanges with Sessions and Fox attest; for references see Fox 1986 and Sylvan 1990. While deep-green theory *is* committed to rational procedures, logical methods and so forth, it is at the same time heavily committed to changing all these things radically (as already glimpsed with satisization displacing maximization in rationality and other directives).

36 There is an advertised problem looming here for Deep Ecology, with, on the one side, its obscure and lax admission criteria and, on the other, its excessively exclusive list of recognized Deep Ecologists.

CHAPTER 6

1 This was written in the 1980s, the decade which flaunted *greed is good,* a theme that has not been substantially abandoned but simply driven underground. It is a theme supported by a rumbustious disgraceful utilitarianism from which utilitarians have been trying, with limited success to distance themselves (by different aggregation and trickle-down assumptions, etc.). Since the late 80s there has been a *weak* ethical revival: ethics is increasingly noted, in word, if not satisfactorily in deed. But the contention in the text still holds.

2 All values must answer ultimately to (perhaps concealed) intrinsic values: see Routley and Routley 1979.

3 We are indebted in these observations to John Patterson.

4 Positions on the other side of a main divide in contemporary philosophy,

empiricism (as opposed especially to transcendentalism) can enjoy considerable mystique, often combining this with verve. Thus Hume, whose position runs into obscurity and inconsistency once its initial bold clarity is left behind. Similarly James, once his 'radical empiricism' merged with pluralism and religion. Similarly Quine, who likewise has gained a certain mystique, including a requisite level of obscurity once one passes beyond the superficial clarity of his pronouncements – as well as enjoying the immense present advantage of being a Northeastern American.

5   While a reductive analysis, analysing rights away is bound to fail, a semantical analysis, which is not reductive, need not. (Similarly, for axiological and other representations.) Compare the situation with that of possibility, where a satisfactory semantical analysis, through worlds, does *not* deliver a syntactical translation.

6   Naess 1988.

7   Some shapers of newer environmentalism have taken different positions. There are firstly those who maintain that those sorts of analyses and assessment procedures are part, an integral part, of the older shallower paradigm, that ought to be left behind (thus e.g. Sessions, also Fox). There are secondly those who assume that such methods are essentially utilitarian (thus e.g. Attfield), not necessarily a shallow doctrine. Both are mistaken. For some elaboration of such analyses on a nonutilitarian value-theoretic basis, see Routley and Routley 1973.

8   E.g. Routley 1984, Routley and Routley 1985.

9   Naess and Rothenberg 1989. Theoretical treatments of improved measures and objectives have been around for several decades; see Routley and Routley 1973.

10   Such observations also help in undermining what remains of the dubious doctrine of comparative advantage. For elaboration, see Routley and Routley 1973.

11   For more detail and argument, see Routley 1984.

12   Naturally not all *frees* so disappear, e.g.. free will, free city, perhaps free rider.

13   The main, still preliminary, work done is that for deep-green theory, for which see Routley and Routley, 1979. But there is much closely related enterprise, e.g. Trainer, Bookchin, and many others.

14   Indeed outside the simplest settings there is *no pure practice.* What looks like 'pure practice' is practice informed by old entrenched theory, much of it pathetically shallow stuff descended from Hobbes, Locke and Co., which has become conventional wisdom (from which moderate balanced people do not diverge). Herein lies a main fallacy in that popular theory of pure practice.

15 A little of which Naess indicates elsewhere. Also see esp. Devall 1988. There is a wealth of individual lifestyle handbooks, containing material concerning or hovering upon deeper lifestyles and individual practice. By comparison, there is little on social practice.

16 Naess 1988, 7.

CHAPTER 7

1 Bennett 1985, 16.

2 Dr. Noel Brown of the United Nations Environment Program, is one. Suzuki, the Ehrlichs, Clark, the Meadows, are among others.

3 Goldsmith and Hildyard 1988, 229. For many such examples of improved environmental performance by companies, see Schmidheiny 1992. Regrettably, such examples are still exceptional, in an industrial world of progressively declining environmental quality.

4 Binswanger 1989, 79.

5 Binswanger 1989, 79. Other macroeconomic targets should be set also. With high inflation investment in land becomes much more attractive than investment in financial instruments.

6 Passmore 1974, 53-4.

7 Passmore 1974, 68.

8 Foreman 1987, 24.

9 For some details of these agendas, details not needed here, see Bowers.

10 Environmental work on education exhibits the same familiar phenomenon. There appears far more good critical material, critical of unsatisfactory, and present, practices, than there is positive material explaining tenets of satisfactory education. Good negative material mostly gives way to positive waffle – vague, evasive and unteachable proposals. See further below.

11 For some critical development of these themes, many of them Enlightenment themes, see Sylvan 1994, Chapter 10.

12 A worthwhile account of how education serves the status quo (and thus increasingly serves business interests) and blocks socio-economic change is presented in Pepper 1984, p.218ff. Stock education operates not merely by omission – failure to encourage critical awareness and ability to think independently (to say the least) – but also by commission through presentation of a 'hidden curriculum' incorporating dominant values and ideological themes (such as capitalism, its needs and preparation for it, nationalism, social Darwinism, positivism, and so on). Though Pepper

246

contends that changes in education will have to be concurrent with other considerable reforms in society, we see no reason why education should not partially anticipate and motivate social change.

For much more on the severe shortcomings in present environmental education, such as it is, and a little on what to do about it, see the special issue of *New Education* on environmental education (vol. xi(2), 1989).

13 This list is drawn from *The Trumpeter* 8:3, and is especially indebted to p. 98. The authors are no doubt right, however, in insisting that deep education is not just about training experts and specialists, as well as about much else.

14 For one introduction to the ideas behind such education, see Sylvan 1994, chapter 3. Ideally some recent history and philosophy of mathematics would also be inserted into the curriculum (it would help too in ending the absolutism and whiz-kid stuff that often still dominates much of the curriculum).

15 On the major roles of biohistory, see Boyden 1992. But the sorts of biohistories would be considerably more diverse and tentative, as we do not share Boyden's confidence regarding mainstream science, its correctness, benignness, and ideological neutrality.

16 The appalling idea, which enjoys a prestigious modern history, is found recently in *Scientific American*; the phrase is quoted in Orr 1991, p. 99, who documents some of the general myths. On myths in forestry, see e.g. Routley and Routley 1973 and Lansky 1992. For investigations countering some of the general myths, see Sylvan 1994, chapters 9 and 10.

17 Adams 1979, 62.

18 Austin 1989, 39.

19 Eckersley 1988, 9.

20 Eckersley 1988, 9.

CHAPTER 8

1 Types of change are explained in book ranging from the splendidly Spartan *Living Poor With Style* of Callenbach 1972 to such lavishly illustrated texts such as Seymour and Girardet's *Blueprint for a Green Planet*. Less comprehensive as regards low-impact lifestyles are green consumerism handbooks.

2 Improved informed voting would be much assisted by well publicized assessments of the environmental performances, environmental statements, and voting patterns of politicians, and of their willingness to

support development interests or to succumb to business pressures and blackmail, e.g. to reiterate exaggerated and misleading claims as to short-term revenue-generating and job-creating prospects of anti-environmental development projects, or alleged job losses if development does not eventuate. Hopefully many more politically engaged environmental groups will emerge to undertake this sort of political activity (already underway in parts of USA).

3 Mendes 1989, 5.
4 Dodd 1989, 8.
5 Dodd 1989, 8, rearranged.
6 Goldsmith and Hildyard 1988, 220.

Bibliography

Abrahamson, D. (1985) "Creatures of Invention", *National Wildlife*, 25: 25-28.

Adams, D. (1979) *The Hitch Hiker's Guide To The Galaxy*. London: Pan.

Ahmad, Y.J. and M. Kassas (1987) *Desertification: Financial Support for the Biosphere*. Sydney: Hodder and Stoughton.

Attfield, R. (1983) *The Ethics of Environmental Concern*. Oxford: Basil Blackwell.

Austin, N. (1988) "Our Zombie Fauna", *Bulletin*, 25 October: 48-52.

Austin, N. (1989) "Our Dying Oceans – Can We Save Them?", *Bulletin*, 24 January: 36-42.

Benn, S.I. (1975) "Personal Freedom and Environmental Ethics: The Moral Inequity of Species", Presented at the World Congress on Philosophy of Law and Social Philosophy, St. Louis, Mo. August, 1975. 34pp.

Bennett, D.H. (1985) "Inter-Species Ethics: A Brief Aboriginal and Non-Aboriginal Comparison". Canberra: Australian National University, Departments of Philosophy (Discussion Papers in Environmental Philosophy, 7).

Bennett, D.H. (1986) "Inter-Species Ethics: Australian Perspectives". Canberra: Australian National University, Departments of Philosophy (Discussion Papers in Environmental Philosophy, 14).

Bennett, D.H. and R. Sylvan (1988) "Ecological Perspectives on an Expanding Human Population", in L.H. Day and D.T. Rowland (eds) *How Many More Australians?* 153-166. Melbourne: Longman Cheshire.

Bentham, J. (1823) *An Introduction to the Principles of Morals and Legislation*, J.H. Burns and H.L.A. Hart, (eds). London: Athlone Press, 1970.

Binswanger, H. (1989) "How Brazil Subsidises the Destruction of the Amazon", *The Economist*, 18 March: 79.

Bodian, S. (1982) "Simple in Means, Rich in Ends: a conversation with Arne Naess", *Ten Directions*, Sum/Fall, 7: 10-12.

Bookchin, M. (1990) *Remaking Society*. Boston: South End.

Bowers, C.A. (1991) "Anthropocentric foundations of educational liberalism", *The Trumpeter* 8(3): 102-107.

Boyden, S. (1992) *Biohistory: the interplay between human society and the biosphere*. Pearl River NY: Parthenon Publishing Group.

Brennan, A. (1991) "Environmental awareness and liberal education", *British Journal of Educational Studies*, 39: 279-298.

249

Bibliography

Burnheim, J. (1985) *Is Democracy Possible?* Cambridge: Polity.

Callenbach, E. (1972) *Living Poor with Style.* New York: Bantam..

Callicott, J.B. (1979) "Elements of an Environmental Ethic: Moral Considerability and the Biotic Community", *Environmental Ethics,* 1: 71-81.

Callicott, J.B. (1982) "Hume's *Is/Ought* Dichotomy and the Relation of Ecology to Leopold's Land Ethic", *Environmental Ethics,* 4: 63-74.

Carson, R. (1965) *Silent Spring.* Hammondsworth: Penguin Books.

Chubb, P. (1988) "Earth and Fire", *Time Australia,* 48: 14-41.

Commission for the Future (1989) *Personal Guide for the Earth,* 1st ed. Canberra: Australian Government Publishing Service.

Crothers, N. (1993) "Cradle to Grave", *Consuming Interest,* June: 4-7.

Daly, H. and J.B. Cobb (1989) *For the Common Good.* Boston: Beacon Press.

Devall, B. (1988) *Simple in Means, Rich in Ends.* Salt Lake City: Peregrine Smith.

Devall, W. (1980) "The Deep Ecology Movement", *Natural Resources Journal,* 20: 299-322.

Devall, W. and G. Sessions (1985) *Deep Ecology: Living As If Nature Mattered.* Salt Lake City: Gibbs M. Smith.

Dobson, A. (1990) *Green Political Thought.* London: Routledge.

Dodd, M. (1989) "Soviet Greenies Face Harassment", *Canberra Times,* 9 January: 8.

Drengson, A.R. (1989) "Reflections on Ecosophy", *Deep Ecologist,* 31: 13-15.

Eckersley, R. (1988) "The Need to Raise Our Eyes to the Horizon of Our Future", *Canberra Times,* 29 November: 9.

Eckersley, R. (1992) *Environmentalism and Political Thought.* Albany NY: State University of New York Press.

Edwards, J. (1982) *Ethics without Philosophy.* Tampa: University Presses of Florida.

Ehrlich, P., A. Ehrlich and J. Holdren (1977) *Ecoscience: population, resources, environment.* San Francisco: W.H. Freeman.

Engel, J.R. and J.B. Engel (1990) *Ethics of Environment and Development.* London: Belhaven Press.

Foreman, B. (1987) "Plant Propagation and Environmental Education", *Environment and Planning,* 6: 24-25.

Foreman, D. and B. Haywood (1987) *Ecodefense: a field guide to monkeywrenching,* 2nd ed. Tucson: Ned Ludd Books.

Fox, W. (1984) "Deep Ecology: a new philosophy of our time?", *The Ecologist,* 14: 194-204.

Bibliography

Fox, W. (1986) *Approaching Deep Ecology*. Centre for Environmental Studies, University of Tasmania.

Fox, W. (1990) *Towards a Transpersonal Ecology*. Boston: Shambala.

Frankel, Sir O. and M. Soulé (1981) *Conservation and Evolution*. Cambridge: Cambridge University Press.

Frankena, W.K. (1979) "Ethics and the Environment", in K.E. Goodpaster and K.M. Sayre (eds) *Ethics and Problems of the 21st Century*, 3-20. Notre Dame: University of Notre Dame Press.

Gever, J. (1991) *Beyond Oil: the threat to food and fuel in the coming decades*, Third ed. Niwot, Colo: University Press Colorado.

Godfrey-Smith, W. (1979) "The Value of Wilderness", *Environmental Ethics*, 1: 309-19.

Godfrey-Smith, W. (1980) "The Rights of Non-humans and Intrinsic Values", in D.S. Mannison, M.A. McRobbie, and R. Routley (eds) *Environmental Philosophy*, 30-47. Canberra: Australian National University, Research School of Social Sciences, Department of Philosophy.

Goldsmith, E. and N. Hildyard, eds (1988) *The Earth Report: monitoring the battle for our environment*. London: Mitchell Beazley.

Gould League of Victoria (1982) *The Edge of Extinction: Australian wildlife at risk*. Prahran, Vic.: Gould League of Victoria.

Greenwald, D., ed-in-chief (1982) *Encyclopaedia of Economics*. New York: McGraw Hill.

Gunn, A. (1986) "What is an Environmental Ethic?", *N.Z. Environment*, 49: 3-7.

Hawke, R.J.L. (1989) *Our Country Our Future*. Canberra: Australian Government Publishing Service.

Kropotkin, P. (1913) *Fields, Factories and Workshops*, second ed. London: Nelson.

Lansky, M. (1992) *Beyond the Beauty Strip*. Gardiner, Maine: Tilbury House.

Leibniz, G. (1846) *Discourse on Metaphysics*, Peter G. Lucas and Leslie Glint, trans. Manchester: Manchester University Press, 1961. (Translation based on diplomatic text of 1907.)

Leopold, A. (1953) *Round River: from the journals of Aldo Leopold*. London: Oxford University Press.

Leopold, A. (1966). *A Sand County Almanac*. New York: Oxford (First published 1949).

Lovins, A. and H. (1987) "Building real security", in T. Woodhouse (ed.) *People and Planet*, 10-21. Devon: Green Books.

McCloskey, H.J. (1965) "Rights", *Philosophical Quarterly*, 15: 115-127.

251

McCloskey, H.J. (1983) *Ecological Ethics and Politics.* Totowa, N.J.: Rowman and Littlefield.

McLaughlin, A. (1993) *Regarding Nature: Industrialism and Deep Ecology.* Albany, NY: State University of New York Press.

Mannison, D.S., M.A. McRobbie and R. Routley, eds (1980) *Environmental Philosophy.* Canberra: Australian National University, Research School of Social Science.

Margalef, R. (1968) *Perspectives in Ecological Theory.* Chicago: University of Chicago Press.

Meadows, D.H. et al. (1972) *Limits to Growth.* New York: Universe Books.

Meadows, D.H. et al. (1992) *Beyond the Limits.* Post Mills Vermont: Chesley Green Pub. Co.

Mendes, O. (1989) "Murder Rife in Rubber Wrangle", *Canberra Times,* 16 April: 5. (Reprint of paper presented by Candido Mendes at "The Amazon and Economic Disorder" Seminar, Rio de Janeiro, October 1988.)

Midgley, M. (1975) "Book review: J. Passmore's *Man's Responsibility for Nature*", *Philosophy,* 50: 106-113.

Moore, G.E. (1903) *Principia Ethica.* Cambridge: Cambridge University Press.

Naess, A. (1973) "The Shallow and The Deep, Long-Range Ecology Movement: a summary", *Inquiry,* 16: 95-100.

Naess, A. (1978) "Excerpt from incomplete translation of *Ekologi, Samhälle och Livsstil* into English". Oslo: unpublished, 7pp.

Naess, A. (1983) "The Deep Ecology movement: some philosophical aspects", *Philosophical Inquiry,* 8: 10-31.

Naess, A. (1984) "What is Basic to Deep Ecology". Canberra: unpublished. 5pp.

Naess, A. (1988) "The Basics of Deep Ecology", *Resurgence,* 126: 4-7.

Naess, A. (n.d.) "Individualism or Collectivism?" Oslo: unpublished, 3pp.

Naess, A. and D. Rothenberg (1989) *Ecology, Community and Lifestyle.* Cambridge: Cambridge University Press.

Naess, A. and G. Sessions (1984) "Basic Principles of Deep Ecology", *Ecophilosophy,* VI: 3-7.

Narveson, J. (1977) "Animal Rights", *Canadian Journal of Philosophy,* 7: 161-178.

Nash, R. (1989) *The Rights of Nature: a history of environmental ethics.* Madison Wis: University of Wisconsin Press.

O'Neill, J. (1993) *Ecology, Policy and Politics.* London: Routledge.

Orr, D. (1991) "What is education for?", *Trumpeter,* 8(3): 99-101.

Bibliography

Ovington, D. (1978) *Australian Endangered Species: mammals, birds and reptiles.* Stanmore, NSW: Cassell Australia.

Passmore, J. (1974) *Man's Responsibility for Nature,* (2nd ed. 1980). London: Duckworth.

Passmore, J. (1975) "Attitudes To Nature", in R.S. Peters (ed.) *Nature and Conduct,* 251-64. London: Macmillan.

Pepper, D. (1984) *The Roots of Modern Environmentalism.* London: Croom Helm.

Rachels, J. (1976) "Do Animals Have the Right to Liberty?" in T.Regan and P. Singer (eds) *Animal Rights and Human Obligations,* 1-20. Englewood Cliffs, NJ: Prentice-Hall.

Reed, C. (1988) "The New Age of Nature's Warriors", *Melbourne Age,* 25 June: Saturday Extra 3.

Regan, T. (1982) *All That Dwells Therein.* Berkeley: University of California Press.

Rivers, P. (1977) *Living Better with Less.* London: Turnstone.

Rodman, J. (1977) "The Liberation of Nature?", *Inquiry,* 20: 83-131.

Roll, E. (1992) *A History of Economic Thought,* Fifth ed. London: Faber & Faber.

Routley, R. (1973) "Is There a Need for a New, an Environmental Ethic?", Varna: Proceedings of the 15th World Congress of Philosophy, 205-210.

Routley, R. (1982) "In Defence of Cannibalism: I". Canberra: Australian National University, Departments of Philosophy (Discussion papers in Environmental Philosophy, 2).

Routley, R. (1984) "Maximizing, Satisficing, Satisizing: The Difference in Real and Rational Behaviour under Rival Paradigms". Canberra: Australian National University, Departments of Philosophy (Discussion papers in Environmental Philosophy, 10).

Routley, V. (1975) "Critical Notice (review of *Man's Responsibility for Nature,* by J. Passmore)", *Australasian Journal of Philosophy,* 53: 171-185.

Routley, R. and V. Plumwood (1984) "Moral dilemmas and the logic of deontic notions" #6; also in *Paraconsistent Logic* (Phil Verlag 1989).

Routley, R. and V. Routley (1973) *The Fight for the Forests* 1st ed. Canberra: Australian National University, Departments of Philosophy.

Routley, R. and V. Routley (1973a) "Semantical foundations of value theory", *Nous,* 17: 441-456.

Routley, R. and V. Routley (1979) "Against the Inevitability of Human Chauvinism", in K.E. Goodpaster and K.M. Sayre (eds) *Ethics and Problems of the 21st Century,* 36-59. Notre Dame: University of Notre Dame Press.

Bibliography

Routley, R. and V. Routley (1980a) "Human Chauvinism and Environmental Ethics", in D.S. Mannison, M.A. McRobbie, and R. Routley (eds)*Environmental Philosophy*, 96-189. Canberra: Australian National University, Research School of Social Science.

Routley, R. and V. Routley (1980b) "Social Theories, Self Management, and Environmental Problems", in D.S. Mannison, M.A. McRobbie, and R. Routley, (eds)*Environmental Philosophy*, 217-332. Canberra: Australian National University, Research School of Social Sciences.

Routley, R. and V. Routley (1985) "An expensive repair kit for utilitarianism". Canberra: Australian National University, Departments of Philosophy (Discussion Papers in Environmental Philosophy, 7).

Sagoff, M. (1988) *The Economy of the Earth*. New York: Cambridge University Press.

Sale, K. (1985) *Dwellers in the Land: The Bioregional Vision*. San Francisco: Sierra Club Books.

Schmidheiny, S. (1992) *Changing Course: a global business perspective on development and the environment*. Cambridge Mass: MIT Press.

Schweitzer, A. (1933). *My Life & Thought*, C.T. Champion, trans. London: George Allen & Unwin.

Seymour, J. and H. Girardet (1987) *Blueprint for a Green Planet: how you can take practical action today to fight pollution*. North Ryde, NSW: Angus and Robertson, in association with Dorling Kindersley.

Sessions, G. (1980) "Shallow and Deep Ecology: a review of the philosophical literature", in R.C. Schultz and J.D. Hughes (eds) *Ecological Consciousness: essays from the Earthday X colloquium*. Washington, D.C.: University Press of America.

Singer, P. (1976) "All Animals Are Equal", in T. Regan and P. Singer (eds) *Animal Rights and Human Obligations*, 148-162. Englewood Cliffs, N.J.: Prentice-Hall.

Singer, P. (1977) *Animal Liberation*. London: Granada.

Singer, P. (1979) "Not For Humans Only: The Place of Nonhumans in Environmental Issues", in K.E. Goodpaster and K.M. Sayre (eds) *Ethics and Problems of the 21st Century*, 191-206. Notre Dame: University of Notre Dame Press.

Singer, P. (1989) "Unkind to Animals", *New York Review of Books*, 2 February: 36-38.

Singer, P., ed. (1991) *A Companion to Ethics*. Oxford: Blackwell.

Smith, A. (1776) *The Wealth of Nations*. London: J.M. Dent (1937-1938).

Bibliography

Southwick, C. (1972) *Ecology and the Quality of Our Environment.* New York: Van Nostrand Reinhold.

Sylvan, R. (1985) "A Critique of Deep Ecology". Canberra: Australian National University, Departments of Philosophy, (Discussion Papers in Environmental Philosophy, 12).

Sylvan, R. (1986) "Three Essayes Upon Deeper Environmental Ethics". Canberra: Australian National University, Departments of Philosophy, (Preprint Series in Environmental Philosophy, 13).

Sylvan, R. (1988) "Deep Ecology and Deep-green Theory". Canberra: unpublished, 4 pp.

Sylvan, R. (1990) "In defence of deep environmental ethics". Canberra: Australian National University, Departments of Philosophy (Preprint Series in Environmental Philosophy, 18).

Sylvan, R. (1992) "Process and action", *Studia Logica,* 51: 379-438.

Sylvan, R. (1992a) "On the value core of deep-green theory", in G. Oddie and R. Perrett (eds) *Justice, Ethics, and New Zealand Society,* 222-229. Auckland: Oxford University Press.

Sylvan, R. (1993) "What *is* wrong with applied ethics", in *Against the Main Stream* (Preprint Series in Environmental Philosophy, 20). Canberra: Australian National University, Philosophy Program.

Sylvan, R. (1993a) "Paradigmatic roots of environmental degradation", in *Against the Main Stream* (Preprint Series in Environmental Philosophy, 20). Canberra: Australian National University, Philosophy Program.

Sylvan, R. (1994) *Deep Plurallism.* Canberra: typescript.

Sylvan, R. (1994a) *Green Anarchism.* Canberra: typescript.

Sylvan, R. and D. Bennett (1990) "Utopias, Tao and Deep Ecology". Canberra: Australian National University, Departments of Philosophy, (Preprint Series in Environmental Philosophy, 19).

Taylor, A-M. (1987) "The Psychology of Activism", in F. Fisher (ed) *Sustaining Gaia: Contributions to Another World View,* 100-104. Clayton, Vic.: Monash University Graduate School of Environmental Science (Papers from: *Environment, Ethics and Ecology II,* 1984).

Townend, C. (1981) *A Voice for the Animals: How Animal Liberation Grew in Australia.* Kenthurst: Kangaroo Press.

Trainer, T. (1985) *Abandon Affluence!* London: Zed Books.

Wells, D., R. Elliot, T. Lynch and W. Grey, "Ethical and political aspects of environmentalism", Course Notebook for Philosophy/Politics 265/365-1. Armidale: Department of Philosophy, University of New England.

255

Wenz, P. (1988) *Environmental Justice.* Albany NY: State University of New York Press.

Wilson, E.O. (1988) *Biodiversity: National Forum on Biodiversity.* Washington DC: National Academy Press.

White, L. Jr. (1967) "The historical roots of our ecologic crisis", *Science* 155: 1203-7.

World Commission On Environment And Development (1987) *Our Common Future.* Oxford: Oxford University Press.

World Resources Institute (1992) *World Resources 1992-93.* New York and Oxford: Oxford University Press.

Index

Foucault, M. 174
Fox, W. 98, 103, 109, 154, 228,
    236, 242, 244, 245
Frankel, Sir O. 238
Frankena, W.K. 234
free good 6
free markets 172
free trade 135
fur coats 215
futurology 194

G

Gandhi 212
GDP 169
gene pool argument 65–66
genocide 118
Gestalt 103, 106–107, 113. *See
    also* value Gestalt
Gever, A. 233
Girardet, H. 210, 247
glasnost, green 214
GNP 129, 131
God 8, 79, 91
Godfrey-Smith, W. 234
Goldsmith, E. 246, 248
goodness 143–144
    as nonjective quality 143
goods 162–163
Gordon-below-Franklin dam 182
Gould League 238
greater value assumption
    88, 90, 93, 139
greeen 229
greeeening 229
green 21–22
Green Ban movement 215
Green Ethical Fields 228
green ethics 24
green ideosystem 24

green platform 23
green representatives 176
Green, T.H. 107
greening of ethics 25
    as historical extension 29–31
    defective accounts of 27
Greenwald, D. 227
Greenway 214
Grey, W. 65, 66, 79
growth
    economic 134
guanays 115
guano 115
Gunn, A. 229
gut reaction 145

H

hamburgers 183, 202
handicap predicates 244
Harvard University 28
Hawke, Bob 242
Hegel, G.W.F. 103
Heidegger 231
Hildyard, N. 248
historical extension thesis 29
Hitch Hiker's Guide to the Galaxy
    197
Hobbes 245
holism 148, 163
    extreme 153
    moderate 154
    total 103–104
holism-partism division 148
holistic metaphysics 102–107
human chauvinism 137, 140
human egoist thesis 149
human interference, excessive 113–
    117
human participation 105